59.50
80E

D0874288

Operations Management
in the
Forest Products
Industry

The first condition of success in any job is not brains but character.
Over and over again I have seen men of moderate intelligence
come to the front because they had courage, integrity, self-respect,
steadiness, perseverance, and confidence in themselves, their cause,
and their work. Which of these qualities comes first it is not easy to say,
but certainly courage, perseverance, and self-respect rank high.

GIFFORD PINCHOT, 1865–1946

*Also published by Miller Freeman Publications
for the forest products industry:*

Plywood Manufacturing Practices, second edition
by Richard F. Baldwin

Quality Control in Lumber Manufacturing
edited by Terence D. Brown

Small Log Sawmills
Profitable Product Selection, Process Design and Operation
by Ed M. Williston

Lumber Manufacturing
The Design and Operation of Sawmills and Planer Mills
by Ed M. Williston

Modern Particleboard and Dry-Process Fiberboard Manufacturing
by Thomas M. Maloney

Dry Land Log Handling and Sorting
Planning, Construction, and Operation of Log Yards
by Charles M. Hampton

Saws: Design, Selection, Operation, Maintenance
by Ed M. Williston

The Logging Business Management Handbook
by Ronald R. Macklin

Logging Practices: Principles of Timber Harvest Systems, revised edition
by Steve Conway

Timber Cutting Practices, third edition
by Steve Conway

Modern Sawmill & Panel Techniques, volumes 1 and 2
Proceedings of the North American Sawmill & Panel Clinics,
Portland, Oregon, 1980 and 1981

Modern Sawmill Techniques, volumes 1 through 9
Proceedings of the North American Sawmill Clinics

Modern Plywood Techniques, volumes 1 through 7
Proceedings of the North American Plywood Clinics

Electronics in the Sawmill
Proceedings of the Electronics Workshop at the Sawmill & Plywood
Clinic, Portland, Oregon, 1979

Business Management for Sawmill Operators
Proceedings of the Business Management Clinic for Sawmill Operators
at the Sawmill & Plywood Clinic, Portland, Oregon, 1979

Sawmill Techniques for Southeast Asia, volumes 1 and 2
Proceedings of the Southeast Asia Sawmill Seminars

Operations Management
in the
Forest Products
Industry

RICHARD F. BALDWIN

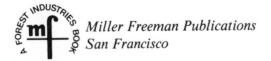 *Miller Freeman Publications*
San Francisco

A FOREST INDUSTRIES BOOK

Copyright © 1984 by Richard F. Baldwin

Library of Congress Catalog Card Number: 84-61889
Hardcover ISBN 0-87930-159-7
Softcover ISBN 0-87930-160-0

Quote on first page from *Breaking New Ground* by Gifford Pinchot
(New York: Harcourt, Brace and Company, 1947), p. 283.

All rights reserved. No part of this book covered by the copyrights
hereon may be reproduced or copied in any manner whatsoever
without written permission except in the case of brief quotations
embodied in articles and reviews. For information contact
the publishers, Miller Freeman Publications, Inc.,
500 Howard Street, San Francisco,
California 94105, USA.

Printed in the United States of America.

Contents

Foreword

The forest products industry is in transition . . . a statement few would refute. Change is inevitable, but its nature is neither easily defined nor predictable. Nonetheless we must be ready and able to cope with it—and we need some basis for the necessary adaptations.

What do transition and change mean to managers, owners and operators of fiber based companies? Without a doubt, they create a feeling of excitement . . . and a sense of anxiety. The anxiety comes from the difficulty of coping with new demands and expectations.

In order to deal with change, it is necessary to be adaptable. It is also necessary to be technically and managerially knowledgeable in today's complex world of competition, advanced technology and rapidly changing economic and environmental conditions. And it helps to have some way of systematically obtaining knowledge. The author of this book has provided just such a way.

The opening chapters, with a description of the heritage of the industry, provide a perspective on today's standards and practices. Even though we may keep abreast of current trends, it is useful to understand our industry's history in order to better forge ahead into the ever changing future. For the experienced mill manager there will be moments of nostalgia when reading these chapters; for the younger manager or owner there will be better understanding of the practices and idiosyncracies of the industry . . . even as practiced today.

Change brings with it the need for planning. Diving headlong into problems without planning may bring results but often without a full perspective and the best potential for success. For this reason, the specific planning methods presented by the author are invaluable. The industry has become complex; planning for and dealing with this complexity is imperative. Planning is a key to improving the success of a company because it forces us to look at alternatives before they occur. By looking at the albeit unpredictable future we can react more quickly to change when it occurs because we have anticipated possible outcomes.

It is no longer an easy task to make the returns on investment that were possible when stumpage was cheap or labor inexpensive. Today's operations, including technology, are costly and must be managed correctly to optimize investment. Understanding the processes, and learning the views of others about the processes, can be enlightening. Baldwin contributes valuable insight into how the fiber conversion process works. He shares his years of knowledge by citing specific experiences, which permits the reader to build upon his or her knowledge and, hopefully, better compete both domestically and abroad. Certainly the production process can be thought of as the "heart" of the operations. Learning more about it increases one's ability to handle the dynamics of a changing industry.

Today, a company's profits are often made at the margin but only with a keen understanding of operations management and control. It is no longer possible to walk into the mill, turn the switch and expect it to go. More control techniques must be practiced, with the manager an integral part. Baldwin explains how to combine sound management techniques with understanding and a sensitivity to what controls are necessary. The knowledge a reader gains from the chapters on management techniques may mean the difference between turning a profit and losing the firm.

The author returns to the theme of change in the final section of the book. This time he is explicit about managing for change—from operating in the lean environment to incorporating advanced technology into the converting operations. Knowing what needs to be done and when to do it is as critical for a forest products manager as it is for an athlete in the heat of competition on a playing field. Understanding change and knowing how to react to it may make the difference between being in business only through the good times or surviving one of our downturns. One of the best ways of moving with change is to share philosophy and experiences with others; the broadening that results makes adaptability possible. Baldwin provides a base for understanding and managing change in this industry.

I fully expect this book will soon take on the appearance of a well worn, dogeared, marked volume on your desk or shelf. This is the highest compliment that can be paid to a book and its author because it is a sign of constant use. I believe that you will find, as I have, that this book will become an active part of your operation's resources. Its insight and readability will find favor with new managers as well as seasoned experts in this exciting industry as it moves further through its transition . . . its change and metamorphosis, which will keep the industry alive.

Kenneth D. Ramsing
Professor of Management and
Associate Director of the Forest
Industries Management Center
Graduate School of Management
University of Oregon, Eugene
October 1984

Preface

Manufacturing forest products profitably has never been easy, and industry trends indicate that it is becoming ever more management and capital intensive. How do the successful firms grow and prosper, and how does the industry in general cope with the cyclic business environment? There is no easy answer; what works well for one firm does little for another.

The answer lies not in the size or expertise of the firm or in ready access to cheap timber and receptive markets. In fact, favorable circumstances such as these frequently set up the firm for future problems or failures.

The answer seems to lie more in attitudes and thought processes than in ready resources and markets. Often the successful entrepreneur chooses to accept scarce and expensive resources combined with limited capital and a hostile business environment as challenges and opportunities rather than as barriers and problems. And there is a difference.

One operator, when interviewed, expressed pessimism for his business prospects in particular and the industry in general. Another operator, his mill located about 40 miles further down the highway using the same log type in the same market area, discussed his performance records and achievements with pride; he related that the previous month had been a record profit month.

Both were operating in the trough of a recent forest products market depression, while the country was in a recession generally. Although both operators recognized the problems, one chose to see them as opportunities, and he prospered doing just that.

This book is illustrated with success stories; it focuses on thoughts and thought processes. Practical, time-tested management concepts are shown in action rather than theorized. The illustrations are straightforward and simple; the text is written in a like fashion. There are a number of extracts from *Plywood Manufacturing Practices*; the author has discovered that there is little difference in the methods for profitably managing the various forest products businesses.

Basic manufacturing skills are growing closer across product lines. Increased automation, new technology and the geographic decentralization of manufacturing facilities are blurring the boundaries of the manufacturing process from product to product. This concept was reinforced when a fiberglass pipe manufacturer in Arkansas called to discuss common problems in technology application.

Technology, labor and management skills and a fairly uniform raw material are becoming common to all forest products manufacturing operations. Only the heritage is still widely divergent. Has the romance and the art gone out of manufacturing? Yes, I guess it has to a large extent.

Rising costs, gut-crunching downturns, apparent raw material shortages and the infusion of modern management and technology have taken their toll. But it's still a pretty good business, with adherents who can't imagine themselves doing anything else. The author is one of those.

The text is intended for a broad audience: from the machine operator to the company president, plus all those folks in between who make positive things happen in the industry. While the book is primarily intended for participants or would-be participants in softwood manufacturing, all or part will be useful to the non-forest industry business person.

The book is organized into six sections. The first provides the reader with an overview of the forest products enterprise; the second describes the management heritage, or roots, of the industry. These sections are the underpinnings for the specifics that follow.

The third section, "Management Planning," outlines the strategic and tactical planning process. It also describes the use of the computer as a planning and management tool. Value management, the conceptual process for milling the softwood tree, is defined and its application as a log allocation tool is explained. The management specifics outlined in this section prepare the reader for the chapters on mill operating practices.

The section entitled "The Production Process" is a hands-on description of the manufacturing process. It describes log and raw material handling; it also details the manufacturing process for each primary product. Chapter 14 describes the state-of-the-art in major electronic innovations such as programmable logic controllers, computerized process controllers and other electronic process control devices.

The fifth section, "Management and Cost Control Techniques," highlights specific methods for extracting additional value, optimizing costs and maximizing floor control. This section describes ways to obtain more from less of the resource while using the valuable people resource to its fullest.

Managing change is the focus of the last four-chapter section. It describes methods and shares experiences in operating a mill in a lean business environment; it also describes the start-up or turnaround situation in detail. Two chapters deal with the application of innovation and new technology in the mill. The entire section deals with change, accelerating cycles of change that have to be anticipated and integrated into the firm to ensure survival and profitability over the long term.

Numerous interviews, mill visits, discussions and an extensive literature search provided the framework for this book. Twenty-five-plus years in the industry as hourly employee, staff worker and manager provide the mechanism for filtering ideas and concepts.

The author appreciates those industry leaders, professors, students and vendors who read and offered comments on the various drafts. A special acknowledgement is given to those individual industry leaders who responded with information, photographs and other editorial needs.

The help and encouragement of Marie Hyde, an able assistant in preparing this book, is gratefully acknowledged. Thanks are also given to the Miller Freeman Publications staff, including Jann Donnenwirth and Judith Brown, and to Beverley DeWitt for their assistance and encouragement in publication of this book. Sterling Platt, a valued friend, was often the sounding board for ideas and concepts; he offered encouragement when the time demands became a barrier to completion.

Betty Baldwin and the Baldwin family . . . a special thanks for the sacrifice that time demands extracted from family activities. It is hoped that the time was well spent; the intent is to provide a forum for thoughtful discussion. The end result may be an industry that looks forward with confidence to the upcoming cycles of change and turbulence.

Dick Baldwin
Camden, Texas
June 1984

Section One
AN OVERVIEW

The forest products industry is like the timber resource, diverse in size and character. The operations manager must understand and meet the challenges of cyclic markets, an evolving timber resource, changing employee expectations and emerging process technologies.

CHAPTER ONE

One
The Forest Products Enterprise: A Study in Change

The forest products enterprise is important; it provides a ready source of building materials and related products, with recent annual shipments valued at about $39 billion (Bilek and Ellefson 1981). The enterprise also provides jobs.

Communities as regionally diverse as Philomath, Oregon; Pineland, Texas; and Passasumkeag, Maine, share part of the more than 700,000 jobs provided nationwide. Supplying building products and jobs is important; a fair return to the owner/shareholder makes all that possible.

Each year the forest products manager is finding the goal of a fair return more elusive. Changing business conditions, gyrating market prices, shifting customer demands and an emerging space-age technology are among the factors shaping the manufacturing environment. No less important is the changing quality and quantity of the timber resource.

THE TIMBER RESOURCE

"Smaller, more diverse, with less desirable characteristics" is a frequently overheard description of today's timber resource. The old-growth Douglas fir, stately white spruce and virgin shortleaf pine are about gone; each is being replaced by a new type of timber resource.

Commented a research forester during a speech: "In the West, the data indicates that the softwood sawtimber quality is declining for all species and across all ownership classes. Preferred species such as Douglas fir and redwood are receiving the greatest impact, and non-public lands would appear to be declining in timber quality faster than the public ownerships" (LaBau and Knight 1978, p. 6). Second-growth Douglas fir, the whitewood species such as spruce, hemlock and the true fir, and other diverse softwood and hardwood species are increasingly taking up the slack in the Pacific Northwest.

The regional resource in the Southeast is described as follows: "One of the most striking contrasts in the Pacific coast and the South is the disparity in tree sizes. . . . Three-fourths of all the softwood volume is in trees fourteen inches DBH and smaller. Conversely, only one-fourth of the Pacific coast softwood volume is in trees fourteen inches DBH and smaller" (Beltz 1979, p. 65).

The conclusion: Both regions will implement intensive regeneration efforts; those efforts will result in larger quantities of short-rotation softwoods available for milling. Previously overlooked pockets of small logs adjacent to urban markets, coupled with the utilization of low-grade hardwoods such as poplar and aspen in eastern Canada and the northern tier of states, will add to the commercial timber base. Log diameters and species will no longer be constraints to manufacturing; the characteristics will merely define the manufacturing process after the end-use market has been identified.

The correlation between the timber resource and the manufacturing process is summed up in the following examples. "One of the important items that any forest products organization has to examine is their forest resource. . . . A well-publicized example is a sawmill in British Columbia which cost $20 million for its infrastructure and went into receivership. . . . They really didn't know what they were going to put into that sawmill" (Dobie 1977, p. 21). "Today in Oregon, a sure way for a company to move from the endangered species list to that of an extinct specimen is to manufacture lumber from $400 stumpage using sawmill equipment designed for $2 and $3 stumpage" (Stoltenberg 1979, p. 6).

THE ENTERPRISE

The forest products enterprise is like the timber resource, diverse in size and character. Modernize, merge or dissolve are increasingly the choices facing the owner/manager. A capital-intensive manufacturing process combined with wild swings in market prices and coupled with increased competition for raw material requires ever larger financial resources. Although the owner/entrepreneur continues to control the majority of the individual enterprises, the trend is toward the larger corporate owner.

The corporate owner usually has available the financial resources to:

- Optimize yield from the overall timber resource and from each stem
- Achieve economy of scale
- Maximize man-hour efficiency
- Standardize and automate the process
- Commit to ongoing state-of-the-art technology to realize the cost and value benefits this technology can provide

The result will be fewer but larger facilities, with the incumbent demands upon the manager.

BUSINESS CONDITIONS AND PRODUCT PRICES

The residential building industry, the key customer for the forest products enterprise, has its cyclic ups and downs. The downs can be characterized as follows: "Housing starts dropped 1.7 percent in September to a seasonally adjusted annual rate of 918,000 after declining a revised 10.8 percent in August to an adjusted

The forest products enterprise provides a ready source of building materials for projects such as the condominiums shown.

934,000 annual rate, the Commerce Department reported this week" (National Forest Products Assn. 1981, p. 3). In 1978, less than three years before the previous statement was made, annual housing starts had been more than 2 million.

Commodity prices for lumber and panels react just as rapidly when demand falters. From a Texas newspaper: "The composite price of lumber, compiled from principal lumber goods and tree species, hit $300 per 1,000 board feet in August 1979. A year ago [September 1980], it sank to $192 and recently languished at $178 . . . The low—$153—came in April 1980. That 1980 spring was bleak for the forest industry" (Connolly 1981).

To quote a *Forest Industry Affairs Letter* of the same period, "As summer heats up, Hollywood could make a horror movie out of the housing market. It's that scary" (Sherman 1981, p. 1).

During a down cycle, a conversation between two or more forest products managers will usually include the following comment almost as a salutation, "It's the worst market I've ever seen!"

This comment contrasts sharply with an up-cycle conversation at an Atlanta airport several years ago:

"Oh, you make plywood," he said. "We are going to do that too."

"Why would you want to do that?" I joked without thinking.

He answered just as quickly: "Because we want to haul our money to the bank every Monday, just like you" (Baldwin 1981, p. x).

The forest industry media also reflect this attitude during up cycles: "The bullish news concerning 1972's construction keeps snowballing so quickly it comes close to being frightening. Yet, day after day and week after week, statistics gather that seemingly ordain that 1972 will be the greatest construction year of them all" (*Crow's* 1972). "Southern Yellow Pine ... Like McDonald's during the lunchcrunch ... Buyers are getting in line, particularly for half-inch" (*Crow's* 1979).

Commented a pioneer industry leader: "The secret: the producer makes it big enough in the good times to carry over the lean times."

"Making it big enough" requires efficient use of the timber resource; it also requires detailed strategic planning, financial capital, and people planning for the manufacturing facility and its employees.

THE MANAGEMENT FUNCTION

This comment from an unpublished 1914 management report for a large resource company could have been written today: "The conclusion of our general examination ... is that the company is losing ground. ... The fundamental cause of this condition lies in the lack of cumulatively increasing efficiency of operation—in other words—costs."

The report author added: "An industry that in years of operation does not devise ways and means of materially reducing cost is not advancing. ... We notice pretty generally a lack of initiative and energy."

The report recommended that management "make every man in the organization realize that he must produce the desired results or get out. ... give each official full authority in his particular province and hold him strictly accountable for results."

Attracting and retaining energetic personnel, securing accountable managers, seeking cost reduction and instituting value-adding efforts, and finding more efficient ways of manufacture continue to be the fundamental activities of a forest products manager.

A noted management scholar has said, "Fundamentals do not change, but the specifics to manage them do change greatly with changes in internal and external conditions" (Drucker 1980, p. 9). The forest products manager is challenged by cyclic markets, an evolving timber resource, a changing technology and the expectations of his employees. He must understand his environment and manage the fundamentals well.

REFERENCES

Baldwin, R. F. 1981. *Plywood Manufacturing Practices*. Rev. 2d ed. San Francisco: Miller Freeman Publications.

Beltz, R. C. 1979. Current and Projected Timber Resource and Wood Use Situations in the South. *Timber Supply: Issues and Options*, p. 65. Proceedings at an FPRS Conference, October 1979, San Franciso, Calif. Madison, Wis.: Forest Products Research Society.

Bilek, E., and Ellefson, P. 1981. Wood-Based Industry: Trends in Selected Structural and Economic Factors Through 1977. *Forest Products Journal* 31 (10): 48 (October).

Connolly, P. 1981. Northwest Reels As Lumber Mills Shut or Cut Back. Associated Press release in *Houston Chronicle*, October 21, sec. 3, p. 1.

Crow's Weekly Plywood Letter. March 3, 1972. Portland, Oreg.: C. C. Crow Publications.

―――. March 9, 1979. Portland, Oreg.: C. C. Crow Publications.

Dobie, J. 1977. Sawmills. *Forest Utilization Symposium Proceedings, Winnipeg, Man., September 28–30, 1977*, p. 21. SP No. 6. Vancouver, B.C.: Forintek Canada.

Drucker, P. F. 1980. *Managing in Turbulent Times*. New York: Harper & Row.

LaBau, V. J., and Knight, H. A. 1978. Historic Trends in the Quality of the Timber Resource Base. *Impact of the Changing Quality of Timber Resources*, p. 6. Proceedings No. P-78-21 of the 1978 Annual Meeting, June 28, 1978, Atlanta, Ga. Madison, Wis.: Forest Products Research Society.

National Forest Products Assn. 1981. Housing Starts Decline Again. *Forest Industries Newletter* 231-LL-42 (October 23):3. Washington, D.C.

Sherman, D. 1981. *Forest Industry Affairs Letter*, vol. 14, no. 10 (June 8). Washington, D.C.

Stoltenberg, C. H. 1979. Timber Supply Options—Symbiosis or Conflict. *Timber Supply: Issues and Options*, p. 6. Proceedings at an FPRS Conference, October 1979, San Francisco, Calif. Madison, Wis.: Forest Products Research Society.

Section Two
MANAGEMENT HERITAGE

The pioneer industry leaders were bold, resourceful and tough, and they knew how to gamble to achieve business gains.

CHAPTER TWO

Two
Industry Roots:
People, Products and Producers

The forest products enterprise is one of the world's oldest business forms; its yield of commodity products includes lumber, paper, plywood, reconstituted panels and other related goods. Its raw material, the tree, is highly variable. So is the process and the people employed.

THE PEOPLE

The pioneer industry leaders were bold, resourceful and tough, and they knew how to gamble to achieve business gains. Some—such as the leaders of Weyerhaeuser Co.—had the vision and foresight to achieve those gains.

Frederick Weyerhaeuser told his fellow investors in 1900 as they formed a company to achieve the long-term stewardship of lands and timber, "This is not for us nor our children. It is for our grandchildren" (Weyerhaeuser 1981, p. 7). George H. Weyerhaeuser, Frederick's great-grandson and the current head of the company, reinforced this vision and foresight when he stated: "We have made a way of life here in walking out to meet the future" (Weyerhaeuser Co. 1975, p. 1).

"Walking out to meet the future" has resulted in a number of industry firsts, including the dedication of the first industrial tree farm in 1941. Another was the organization of the first forest research program within an industrial company.

Other industry pioneers were savvy salesmen. Lawrence Ottinger, the founder of the Building Products Division of Champion Intl., has been described as a born salesman with a magic touch that was complemented by his persuasive manner. Ottinger used these talents with his mother; he borrowed $500 from her and went into business.

"When hostilities ceased in Europe, the government found itself with warehouses full of plywood it could not use. Ottinger entered the plywood field as a middleman, taking over government panels and reselling them abroad at a substantial profit. The buyers' strike of 1920 swept the nation. Pickets paraded in front of many firms protesting the high cost of living. Plywood dropped from seventeen cents to six cents a square foot and wasn't moving. Ottinger had money and bought all the plywood he could get his hands on. The price rose and Ottinger pocketed a tidy profit. Ottinger's tiny U.S. Plywood thus emerged as a big factor in the growing industry" (Cour 1955, p. 33).

Henry McCleary was another pioneer. "McCleary had opened a sawmill in 1896 at McCleary with Wheeler-Osgood owning half the stock. A dynamic, driving fellow, Henry McCleary was a prototype of the legendary American 20th Century industrialist. He came west from Ohio after first stopping in Montana to learn how to handle a brace of six guns. Set up the five of spades on a wall and McCleary could put holes in the four corners and center as fast as he could pull the trigger. He often said he never felt fully dressed without his six guns" (Cour 1955, p. 24).

Henry McCleary later constructed a plywood mill in the town he modestly named for himself. This mill, part of the McCleary Timber Co. complex, was built within a few years of softwood plywood's inception.

And then there was a Texas fellow named Jim Ben Edens. Commented W. A. Franke, the president and chief executive officer of Southwest Forest Industries: "Many of you are familiar with Southwest, which started in Arizona with lumber mill and railway in 1917. . . . in the early 1950's . . . Jim Ben Edens moved to Phoenix from Texas . . . acquired control of what was then known as Southwest Lumber Mills.

"Edens was a remarkable man. . . . he had a very high energy level, a good knowledge of what made people tick, and an ability to be very convincing. . . . the company had annual sales of approximately fifteen million dollars and no real earnings. . . . Edens convinced a number of long term lenders to loan him forty million dollars to build our Snowflake mill. . . . it would use ponderosa pine as a principal raw material—a fiber which had not been successfully used in the paper-making process prior to that time" (Franke 1979, pp. 1–2).

Weyerhaeuser, Ottinger, McCleary and Edens were just a few of the many. Colorful leaders continue to pioneer an evolving industry.

PRODUCTS AND PROCESSES

Man early discovered that the log is a good building material; he then discovered that sawing the log into lumber produces an even more useful material. Thus the advent of the forest products enterprise. Pulping wood and processing it into paper products came next; then came the peeling and slicing of logs into veneer and plywood. Lumber, paper and plywood then set the stage for a myriad of other products, among them reconstituted board.

Lumber

The mechanized sawmill, a saw mounted in a machine and powered by something other than human muscle, was developed beginning about 1200 AD. Norway was building sawmills in 1500 AD, and Sweden followed in the next century. The first North American sawmill was built about 1623 or 1624; subsequently, the sawmill followed the frontier as pioneer settlement swept across North America.

Forest products were produced from large logs until after World War II.

There were about 2,541 North American mills in 1810; there were about 21,000 in 1860. The typical mill produced about 4,000 board measure of rough lumber per day. That production provided the means for the transition from the log or sod structure to the frame building.

For example, the more prosperous settlers of the Oregon Territory began replacing their log cabins with homes and businesses constructed of sawn lumber. By 1849 thirty sawmills were scattered throughout what is now the state of Oregon. The mills were concentrated along the lower Willamette and Columbia rivers, where water for power and transportation was abundant.

The industry became more widespread as demand increased. Other entrepreneurs, such as Andrew Pope and William Talbot, located their mills close to water and logs. "First, they [the founders of the present Pope & Talbot] saw that lumber shipped by Pope's relatives in Boston could not arrive fast enough, cheaply enough, or in great enough quantities to meet the enormous potential of the San Francisco market. . . . Their mill had, therefore, to be at tidewater at a point where sailing vessels would have a good anchorage in a safe harbor close by the log source. . . . the most desirable such area was on Puget Sound in the Oregon Territory" (Coman and Gibbs 1978, p. 4). Pope, Talbot, Charles Foster and J. P. Keller became co-partners and subsequently formed the Puget Mill Co., which would become Pope & Talbot. The first mill, Port Gamble, was producing about 15,000 bd ft a day by 1854.

Technological improvements and innovation characterized that mill as well as others, as the tide of the lumber industry followed the settlements from Maine to Minnesota, then from the upper Midwest to the Pacific Northwest and then into the Southeast. Old-growth trees were replaced with second growth; natural stands gave way to the managed forest of today.

Several lumber operations were sawing a million bd ft and more per operation-day by 1900. The mill equipment, largely steam powered, was solid, well manufactured and cut relatively accurate sizes. Unfortunately the years between 1880 and 1950 produced little change in sawmilling technology. It was not uncommon in the early 1950s to manufacture lumber from turn-of-the century equipment. Cheap and abundant logs combined with a "boom or bust" market demand gave little incentive for change.

A diminishing old-growth reserve of prime logs, sharply increased demand and bright overall business prospects finally prompted a more efficient conversion process. The Scandinavians began developing small-log handling, sorting and processing systems in the 1950s. The small-log chipping canter was introduced into North American mills in the next decade.

In addition, sophisticated electronics were introduced. Hiram Hallock describes the event: "Richard Davis, president of the then-small, struggling Unico Co., came to our Laboratory. He brought along with him a small model D.C. electro-servo unit with a digital readout. With this unit it was, as I remember it, possible to divide a single rotation into 5000 equal parts. It was then possible to select any one of these positions with the controller and the output unit would rotate to this position. Feedback monitored the rotational position and corrected any error caused by

Little change occurred in the typical large-log mill for several decades into the twentieth century. (Courtesy Weyerhaeuser Co. Archives)

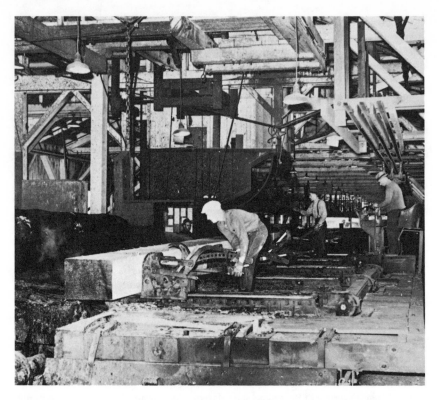

The early mills were labor intensive. (Courtesy Weyerhaeuser Co. Archives)

torque load. . . . I said, 'Dick, this has the makings of the ideal setworks.' He took me seriously and went to the Filer & Stowell Co." (Hallock 1979, p. 13).

With the aid of innovative electronics and computerized controls, the industry has made more and better lumber from progressively smaller diameter logs. U.S. softwood lumber production climbed to about 31 billion bd ft in 1973. Hardwood lumber accounted for an additional 7 billion in that year (Pease 1983, p. 19).

Softwood Plywood

"Stronger than steel pound for pound" is the customary sales pitch given with numerous variations over the years in the effort to promote the structural properties of softwood plywood.

A book commemorating the first 50 years of the industry describes its beginnings: "Sometime in March a panel crew was organized consisting of about six men and supervised by Bailey and Carlson. The plant had a St. Joe lathe and some excellent peeler logs. There was no regular press, no prepared glue, no clipper and no sanding machine in those days. An ordinary steam kiln dryer was the only other piece of equipment used" (Cour 1955, pp. 13–14).

The resulting panels, painstakingly constructed by hand, were placed on exhibit at the Lewis & Clark World's Fair in Portland, Oregon, in June 1905. Thomas Autzen, son of the co-owner of the mill, was in charge of the exhibit and actively promoted the new product. Orders for door skin stock and other products soon followed. In many cases the product sold itself as its unique strength became known.

A local western Washington newspaper cited a need fulfilled: "Before plywood panels, the doors of one McCleary hotel were subjected to a heavy battering, particularly on weekends. . . . This was particularly true of the upper panels, at shoulder height on the doors. . . . The hotel management got wise and installed plywood panels. . . . there were many bruised and battered fists around McCleary until word got around to the celebrants that the new panels would not break, splinter nor crack" (Cour 1955, pp. 28–29).

Door manufacturers were the principal customers for a time. During the 1920s, fir plywood running boards, floor boards and other sheet uses became standard for cars manufactured by Chrysler, Studebaker and Ford, among others. With these and other uses the industry's future seemed secure, but markets shrank quickly when the specified end use required exposure to moisture or weather. As users became more vocal in their delamination complaints, the industry faced a dilemma.

Animal glues, casein, soybean and other glues were tried in sequence. Each offered some improvement but none was waterproof. Finally in 1934 Dr. James Nevin, chief of research at Harbor Plywood, and his staff devised a phenolic-based glue. "The first panels went into the [hot]press at 330 degrees Fahrenheit and Harbor officials held their breaths. When they [the panels] were taken out, the plywood looked good. But this was just the beginning. Tests had to be made. The panels were boiled in water for seven hours a day and for fifteen days" (Cour 1955, p. 91).

The harsh soaking, boiling, freezing, drying and other tests demonstrated that the bond was as strong or stronger than the wood itself. Since that time, the industry product volume has grown in multiples.

About 480 million ft were produced in the USA in 1935. This increased to 1.2 billion in 1945. The industry reached almost 12.5 billion in 1965 and increased upward to nearly 20 billion ft in 1978 and repeated that performance in 1983 (Pease 1984, p. 24).

Small-log peeling technology, developed in the 1950s in central British Columbia, spread to the Intermountain area of Idaho and Montana. The southeastern USA developed an industry beginning in 1963, and this region currently produces about half of the total U.S. softwood plywood produced. Small-log peeling methods plus advanced gluing technology have paved the way for further expansion in other areas and other countries.

Other Panel Products

Particleboard, hardboard, fiberboard and waferboard are among the other major wood products. While lumber meets a basic need and plywood improves upon nature's properties, there is a need for products with additional unique features to serve a much broader market. Furniture components, cabinet flat stock and specialty cut-up uses are among those needs. In addition, reconstituted board products use residual wood from lumber and plywood manufacture and lower value species as raw material.

Particleboard, first developed in Europe, gained in popularity in the 1960s for interior applications such as furniture core stock and underlayment. It displaced both lumber and plywood in a number of applications. Another board, hardboard, demonstrated rapid growth in the 1960s and 1970s. This board, developed in the early 1930s, is used as siding, pegboard for the over-the-shoulder trade, and a substrate for decorative sheet stock.

Medium-density fiberboard, a board that competes with both plywood and particleboard, has excellent edge-machining characteristics. It was introduced in the mid-1970s and became accepted widely in the furniture trade. At about the same time waferboard entered the market as a competitor to plywood sheathing.

Waferboard became economically attractive as the cost of softwood veneer soared between 1970 and 1980. Waferboard production facilities, currently centered in the U.S. upper Midwest and Northeast and in the eastern provinces of Canada, are close to major population centers. Its low-cost furnish is milled from aspen, poplar and other abundant, lower value species. Production of waferboard and its cousin, the oriented-strand board (OSB), is expected to increase sharply. Industry capacity is expected to exceed 3 billion ft^2 (3/8th basis) by the mid-1980s.

THE PRODUCERS

Forest products companies come in all sizes and product orientations. They range from the corporate conglomerate to the single-owner firm. Sales in 1980 ranged from $5.1 million for the largest firm (Georgia-Pacific) to $350,000 or less for the single-mill firm. Capital requirements frequently are small compared with those for other modern industries. Entry barriers are relatively low, but these barriers are expected to increase as the industry becomes more capital intensive. Entry costs will climb sharply as control over timberlands increases in importance.

Three means of entry are common to the majority of the firms: (1) entrepreneur or partnership entry, (2) organized entry and (3) diversification.

Entrepreneur or Partnership Entrants

Entrepreneur or partnership entrants usually begin as loggers, manufacturers or sales firms. For example, Simpson Timber Co. (then known as Simpson Logging Co.) began as a contract railroad builder for a timber company. Sol Simpson, the founder, subsequently began logging under contract as an extension of his railroad construction activities. Contract logging led to independent logging. The latter led to land ownership. Land ownership evolved into manufacturing as a method to provide a return on the difficult-to-market whitewood species developed from company lands.

Roseburg Lumber Co., a large West Coast firm, is another example. Kenneth Ford, the founder, constructed a sawmill near Roseburg, Oregon, in 1936 and installed secondhand machinery. Logging and further integration into plywood and particleboard soon followed.

Georgia-Pacific used a small southern lumber sales firm as a gateway into manufacturing. Owen R. Cheatham, the founder, started the firm in 1927 with the acquisition of a lumber yard in a small Georgia community. Cheatham subsequently weathered the Great Depression while preparing for future growth.

"Survival was possible in part because Cheatham had low overhead and few fixed expenses. He kept inventories low by selling the lumber he purchased from sawmills as rapidly as possible. Sometimes he sold an order of lumber and then purchased the lumber from local mills to fill the order," commented a company-sponsored biographer (Ross 1980, p. 4). As the depression waned, Cheatham continued to develop investor confidence and began to expand. The company owned five southern sawmills by 1938.

During World War II, Georgia-Pacific, then known as Georgia Hardwood Lumber Co., became a large supplier to the armed forces. A Portland, Oregon, buying office was established in 1946 as an outgrowth of the wartime activity. "The northwest in the postwar period was the last frontier for the lumber industry, excluding Alaska. In western Oregon and Washington grew the lush forest of Douglas fir that constituted forty percent of the remaining raw timber in the country" (Ross 1980, p. 29). Cheatham recognized the opportunity and subsequently shifted the company to the Pacific Northwest.

Georgia-Pacific, The Growth Company, is a hard-driving, aggressive company. Within 30 years after its shift to the Pacific Northwest, the company has reached industry leadership; 1980 sales were $5.1 billion, the largest industry figure for that year. A vast network of mills and facilities combined with 4.1 million acres of fee timber makes this company a challenging competitor. "In the end the skeptics paused to study the methods and techniques of Georgia-Pacific's management and to make the maverick a model for the industry" (Ross 1980, p. *xvi*). Ever alert to changing conditions, in 1982 management moved corporate headquarters to Atlanta, Georgia, in order to be in the center of the action in the fast-developing southern pine region.

Simpson Timber Co., Roseburg Lumber Co. and Georgia-Pacific Corp. are just a few of the many successful examples of entrepreneur entry. Another method—the organized entry—has been used by a lesser number—but an extraordinarily successful handful—of companies.

Organized Entrants

International Paper Co., a close number two in 1980 total sales, is within the top five in solid wood product revenue. "In 1898 the International Paper Co. (IP) was organized by combining twenty [paper] mills in New York, Massachusetts, Maine, Vermont and New Hampshire. Its daily production of more than one thousand tons of newsprint and printing paper was backed by extensive timberlands. At one point IP claimed control of three-quarters of American newsprint production" (Amigo and Neuffer 1980, p. 18).

The years prior to the company's formation were perhaps the most difficult before or since for paper manufacturers. The industry had shifted from rags to wood furnish, a raw material that had low production costs. Increased production resulted in massive oversupply and savage price competition. The consolidation of the 20 small, independent operators into one company was a successful attempt to establish an economy of scale in procuring raw materials, establishing production runs and marketing the resulting production.

International Paper has expanded over the years, combining both horizontal and vertical integration in forest products. Diversification outside the core business has also occurred, until the combined sales in 1980 hit $5 billion. In the same year domestic fee timber acreage exceeded 8 million acres, and the company held harvesting rights to an additional 15.6 million acres.

Weyerhaeuser Co. is another example of an organized entry. Frederick Weyerhaeuser, a midwestern mill owner with 43 years of lumber manufacturing experience, led a group of investors in purchasing 900,000 acres of prime timberlands. These lands, located in Washington State, were part of the Northern Pacific Land Grant lands secured from the federal government some years earlier.

Weyerhaeuser and his partners subsequently added sawmills to process the logs from these lands. Land purchases continued over the years. The resulting 6 million acres of fee timber is perhaps the best stand of privately managed timber in the USA, if not in the world. The company's 1980 total wood-based sales, the raw material procured largely from fee lands, were about $4.5 billion. This company currently is ranked within the top 20% of the nation's Fortune 500 companies in sales and number three within the forest products industry. A recent study states: "Either International Paper or Weyerhaeuser could individually qualify as the 42nd largest state, equivalent in size to Maryland but consisting entirely of valuable timberland" (O'Laughlin and Ellefson 1981a, pp. 60–61).

Louisiana-Pacific Corp., organized in 1972 as a spinoff from Georgia-Pacific, is yet another example of an organized entry. In 1971 the Federal Trade Commission issued a complaint citing Georgia-Pacific's growing dominance of the forest products markets. Robert Pamplin, then chairman of the board and chief executive officer of the company, proposed a compromise.

> Pamplin saw little reason to waste corporate funds fighting the agency and also damaging the corporate image and the price of its securities. So he and his fellow officers conceived of a novel solution to the complaint. He proposed to create a new corporation and spin-off part of G-P's assets and properties, a solution the FTC accepted. So Louisiana-Pacific Corporation was blessed in the beginning with $305 million in assets, including the plywood plants and sawmills in Louisiana and Texas; the redwood and pulp operations in Samoa; the Ukiah operations in northern California; the Intermountain operations in eastern Oregon and Washington and Idaho; the pulp and lumber operations in Ketchikan, Alaska; and a window and door operation in Barberton, Ohio. The spin-off satisfied the FTC (Ross 1980, p. 237).

William Hunt, then president of Georgia-Pacific, became the chairman of the board for Louisiana-Pacific. Harry Merlo was selected as president and chief executive officer. The firm has grown in a dynamic fashion. Assets increased 241% over the original amount within 10 years. Production and sales have centered on lumber and panel products, although pulp and paper–related sales are significant. Current sales are in the billion-plus category.

Diversified Entrants

The diversified entrants are many and varied. Some enter the industry to answer a basic need, such as packaging for consumer products. Others enter in a desire to reduce overall risk in a core business. Some have sought a profitable business opportunity that offers growth. Each entrant has had its own reasons, which satisfy the value judgment of its top management.

"Here today and gone tomorrow" is a theme that frequently characterizes these ventures. Such Fortune 500 giants as Time (Temple-Inland, Georgia Kraft), Gulf & Western (Brown Co.), American Can, and ITT (Rayonier) have elected to enter the industry and have also chosen to leave it.

Other companies, such as Proctor & Gamble (Buckeye Cellulose and others) and Mobil (Container Corp. of America), are growing or have announced intentions to grow. Some companies are fairly recent entrants; others, such as Santa Fe Industries, have been in the business for almost 50 years.

Santa Fe became a controlling stockholder in Kirby Forest Industries (then Kirby Lumber Co.) when that company reorganized in 1936. Its founder, John Henry Kirby, recognized the need to own timber and to build cost-effective facilities; he also recognized that his limited resources were insufficient for the tasks ahead. Kirby fought a party to assume controlling interest, and later Kirby Lumber became a wholly owned subsidiary of Santa Fe.

The entrepreneurs, the organized entrants, and the diversified entrants have provided the engine for industry growth. Only a few representative examples have been mentioned. According to the 1977 Census of Manufacturers there are more than 30,000 companies in the solid wood portion of the forest products industry. Another 4,000 firms are listed as part of the pulp and paper industry.

REFERENCES

Amigo, E., and Neuffer, M. 1980. *Beyond the Adirondacks: The Story of St. Regis Paper Company.* Westport, Conn.: Greenwood Press.

Annual Lumber Review & Buyers' Guide (ALRBG). 1981. The Top 100 Lumber Producers: U.S. and Canada, 1980. *Forest Industries* 108(6):7 (May 30).

Coman, E. R., Jr., and Gibbs, H. M. 1978. *Time, Tide and Timber: A Century of Pope & Talbot.* Palo Alto, Calif.: Stanford University Press.

Cour, R. M. 1955. *The Plywood Age.* Tacoma, Wash.: Douglas Fir Plywood Assn.

Douglas Fir Plywood Assn. 1948. *Technical Data on Douglas Fir Plywood for Engineers and Architects.* Tacoma, Wash.

Evans Products Co. 1982. *1981 Corporate Profile.* Portland, Oreg.

Franke, W. A. 1979. A Presentation to the Phoenix Society of Financial Analysts, Phoenix, Ariz., February 21, 1979. Reprint.

Hallock, H. 1979. Sawmilling Roots. *Electronics in the Sawmill,* p. 7. Proceedings of the Electronics Workshop, Sawmill and Plywood Clinic, Portland, Oreg., March 1978. San Francisco: Miller Freeman Publications.

Industry Week. 1981. Financial Analysis of Industry 1980. Vol. 208, no. 6.

Maloney, T. M. 1977. *Modern Particleboard & Dry-Process Fiberboard Manufacturing.* San Francisco: Miller Freeman Publications.

O'Laughlin, J., and Ellefson, P. B. 1981a. U.S. Wood-Based Industry Structure. Part 1: Top 40 Companies. *Forest Products Journal* 31(10):55 (October).

———. 1981b. U.S. Wood-Based Industry Structure. Part 2: New Diversified Entrants. *Forest Products Journal* 31(11):26.

Pease, D. A. 1983. Lumber Producers Hail '83 Rebound After Terrible 1982. 1983 Annual Lumber Review. *Forest Industries* 110(7):19 (July).

———. 1984. Structural Panels, MDF Set Production Records. 1984 Panel Review. *Forest Industries* 111(4):24 (April).

Reed, W. G., with Maunder, E. R. 1977. *Four Generations of Management: The Simpson-Reed Story*. Santa Cruz, Calif.: Forest History Society.

Roseburg Lumber Co. 1976. *Roseburg Woodsman*. A Special Fortieth Anniversary Issue of a Company Periodical. Vol. 22, no. 3 (March).

Ross, J. R. 1980. *Maverick: The Story of Georgia-Pacific*. Portland, Oreg.: Georgia-Pacific Corp.

Western Wood Products Assn. 1981. *1980 Statistical Yearbook of the Western Lumber Industry*. Portland, Oreg.

Weyerhaeuser, G. H. 1981. Forest for the Future: The Weyerhaeuser Story. An address delivered at a national meeting of the Newcomen Society of North America, New York, N.Y., February 12, 1981.

Weyerhaeuser Co. 1975. *Innovations and Trees—Weyerhaeuser: 1900–1975*. Tacoma, Wash.

Whitman, L. A. 1981. *Softwood Plywood Production Statistics*. Management Bulletin No. FA-210. April. Tacoma, Wash.: American Plywood Assn.

Williston, E. M. 1976. *Lumber Manufacturing: The Design and Operation of Sawmills and Planer Mills*. San Francisco: Miller Freeman Publications.

Three
Evolution of the Enterprise: Profile for Success

Forest products entrepreneurs and managers come in three types: the Doers, the Watchers and the Wonderers. The success of the forest products enterprise is in direct proportion to their activities; their activities mirror the firm.

The Doers accept change and make it happen; the Watchers observe and analyze events. The Wonderers usually end up on the fringes; so do their respective companies.

The history of the industry reflects the activities of each of these types; a firm and its leaders may change from one of the approaches to another during certain periods. Frequently, a company will perceive itself as one type when in reality it is another. The low point of a statement-busting market cycle usually provides a period of self-examination; the Doers are the survivors.

INDUSTRY DEVELOPMENT TIME FRAMES

The survivors, the leaders and their respective companies, have emerged from one or more of the four forest products industry historical time periods.

Period One: 1850–1940

The period from 1850 to just before World War II encompassed the major settlement and development of North America and concluded as the industry and the nation emerged from the Great Depression. In between were frequent regional and industry-wide boom and bust periods and throughout, plenty of cheap timber.

For example, the Gold Rush provided a boom market; the uncertainty as the nation approached the Civil War produced the bust. "Common boards had brought as high as $300 or more in 1849–50. By 1855 the bottom had fallen out of lumber prices. The partners were satisfied with $15 to $17 per thousand for common lumber at the mill" (Coman and Gibbs 1978, p. 13).

The paper industry too had its boom and bust periods: "Newspaper circulation increased dramatically during the 1890's ... The Spanish-American War established the sensational front-page headline as a circulation builder" (Amigo and

Neuffer 1980, p. 16). One newspaper, the New York *World*, witnessed its paid circulation jump from 370,000 in 1896 to a record high 1 million in 1898. The bust soon followed.

The conclusion of the Spanish-American War, the resulting drop in newspaper circulation, and the newsprint industry's overproduction combined to provide huge unsold inventories of newsprint. Newsprint was without a price for a time.

The periods described plus the ones that followed weeded out and stabilized a fragmented industry. Those companies that emerged intact had a strong determination of their core business and well-grounded operating and financial expertise in the forest products industry.

Period Two: 1941–1965

Three key events occurred during the mid–twentieth century. First, the industry strengthened its fee timberland base or acquired one. Secondly, the industry integrated and diversified into more products. Third, the industry began to accept radically new technology, such as sophisticated hardware and computers to run that hardware.

Big mills and large logs were common during Period One. (Courtesy Weyerhaeuser Co. Archives)

The trend developed as the industry recognized that:

- Forest resources were not endless. The cost for timberlands, trees and logs could only go up.
- The science of forestry could provide an endless crop from managed lands.
- The carrying costs for lands and timber were high and had to be offset by more products from the log.

An industry leader summed up this period in the industry's history: "We should recognize that movement from timber extraction toward the holding and custodial management of timber on a permanent basis had set some significant forces in motion.

"The carrying costs of such inventories were high—too high to allow the industry to continue to use only a small percentage of one or two species of the trees in each stand, for conversion into a single product. The search began for ways to use "minor" species of wood, and for ways to use those parts of the tree that were wasted in existing sawmilling processes" (Weyerhaeuser 1981, p. 16).

Period Three: 1966–1975

The third time period was the "go-go" era. Forest products firms merged, expanded, acquired and otherwise grew, sometimes in a haphazard fashion. Many chief operating officers and their top managers found themselves unwelcome guests in strange, unrelated businesses. The emphasis on growth and diversification was great. The results were frequently something else than great.

Rather than rags to riches, the business diversification results were from rugs to real estate. Cabinets, resorts, hardware, furniture and equipment manufacturing were just some of the unlikely business deals. Some corporate historians would just as soon erase the whole era. The goal of diversification—to improve and stabilize earnings—was frequently forgotten during the ensuing struggle to save the parent company.

Non-forest products companies also caught the fever. International Telephone & Telegraph acquired Rayonier in 1968; Mobil entered the forest products industry in 1974. These companies plus others have achieved mixed results.

Although often obscured by all this activity, significant state-of-the-art changes were also being implemented. The computer found the sawmill; the plywood layup process was at last automated.

Period Four: 1976–1990

The fourth period is characterized by the following activities.

Divestiture of Nonrelated Businesses. Nonrelated businesses are being divested by both forest products and non-forest products companies. The forest products firms have redeployed or are in the process of redeploying assets from nonrelated fields. These nonrelated fields either require an inordinate amount of senior management attention and company cash or are simply unprofitable for the present owner.

Fewer frills and ever smaller logs are characteristic of the newer computerized mills that mark Period Four.

Consolidation Back to the Core Business. Companies are adhering to a much narrower perception of the operating bounds. They are building on the strength of the core business rather than diversifying simply as a vehicle to increase the stream of quarterly earnings and/or to smooth out the earnings by moving into countercyclic businesses. The management then focuses on the competitive strengths developed within the core business over prior years.

Expansion and Upgrading of the Core Business. The goal is to expand and upgrade the core business; the result is a larger commitment to single-product lines.

For example, in 1970 the top 100 North American lumber producers represented 430 mills that produced 20.4 billion bd ft. In 1980 the top 100 producers owned 641 mills and produced 29.1 billion ft. (See Table 3.1.)

Emphasis on Extracting Maximum Value from the Fee Timber Resource. Mergers and consolidations are vehicles for resource-value maximization.

Table 3.1. The Top 100: U.S. and Canadian Lumber Producers

Year	Mill count	Annual volume (MMBM)	Average per mill (MMBM)
1970	430	20,443	47.5
1980	641	29,059	45.1

Sources: Pease 1971, p. 13; Lambert 1981, p. 6

THE SUCCESSFUL ENTERPRISE: COMMON
INDUSTRY CHARACTERISTICS

Certain characteristics shared by the successful forest products firms have emerged from the four periods. While these characteristics are most visible in the large corporate operation, each holds equally true for the small, successful family operation. As observed from a historical perspective, the distinguishing features are

- Dynamic, purposeful leadership with a well-defined business strategy that includes recognition of acceptable risk.
- Well-managed fundamentals; both financial and operating basics. The specifics are redefined periodically.
- Cost-effective state-of-the-art conversion facilities. These facilities are modernized, built or abandoned in a timely fashion.
- Acquisition of timberlands and management of the trees as a crop. This includes fee ownership or long-term control of timber and timberland.
- A well-defined core business. All business activities are a logical extension of the core business.
- Integration and diversification within a stream of logical and prudent product decisions.
- A marketing organization which is created to match the needs of the parent business.
- An attempt to obtain a significant portion of the product market. For some, this has meant rough plywood sheathing; for others, specialized items such as dimension lumber or export flitches.

Let's review the individual characteristics in detail.

Dynamic Leadership and Well-Defined Strategy

Georgia-Pacific Corp. combined purposeful leadership with a well-defined business strategy. By the mid-1940s this firm had been a seller and producer in the Southeast for two decades.

"In 1947 Cheatham [G-P's founder] broke with his past. . . . His visits to the Northwest and the market for plywood convinced him that he should be in the business of manufacturing plywood. At the time, the post-war demand for housing and other types of construction had created an insatiable market for plywood. He was confident that the market for plywood would grow and wood would continue to have unlimited uses" (Ross 1980, p. 29).

Cheatham subsequently went public with his closely held company, purchased Bellingham Plywood and prepared for further growth. Building ties with banks and financial institutions plus recruiting key managers were the principal preparation activities.

In 1948 the company purchased Washington Veneer at Olympia, Washington, and also acquired a controlling interest in a Springfield, Oregon, plywood mill. Additional mill properties and timberlands were acquired by Georgia-Pacific in the following years.

By 1960 the company had vast holdings in the Pacific Northwest. Cheatham was still convinced that plywood offered other profitable opportunities. Cheatham and his chief operating officer, Robert Pamplin, recognized that further major growth opportunities would have to occur in the South. They then pioneered the southern pine plywood industry.

Georgia-Pacific's Fordyce, Arkansas, plant opened in 1964; Crossett, in 1965; and three additional plants, in 1966. About 20 new plants were build in a 15-year period.

Georgia-Pacific's 17 operating units in the South produced 40.8% of the region's total pine plywood in 1980. It took 48 competitive plants to produce the remaining 59.2%. Nationally Georgia-Pacific produced 18.3% of the U.S. total. And the pace goes on. In 1982, Georgia-Pacific completed the relocation of its headquarters to Atlanta. By that year, the company was producing 7 times as much plywood and 10 times as much lumber from southern pine as it was from northwestern Douglas fir.

"Despite all the rhetoric, we believe this can be a decade of growth. The leaders, however, will be the efficient . . . the integrated . . . the prepared and the well managed. And we plan to remain that way," stated Robert Floweree, then the company's chief executive officer (Georgia-Pacific Corp. 1981, p. 1).

Other individuals have exerted similar leadership within their firms. The common thread: a purposeful leadership that demonstrates the ability to carry out a clearly defined strategy.

Well-Managed Fundamentals

A forest products firm must manage the fundamentals; the industry has an inherent turbulent demand and market structure.

"In turbulent times, an enterprise has to be managed both to withstand sudden blows and to avail itself of sudden unexpected opportunities. This means that in turbulent times the fundamentals have to be managed, and managed well" (Drucker 1980, p. 9).

The fundamentals are both the financial and the operating basics. One of the key financial fundamentals is the maintenance of sufficient liquidity. "In turbulent times, liquidity is more important than earnings" (Drucker 1980, p. 13).

The successful forest products firms have recognized this and have maintained a careful balance between liquidity and taking advantage of opportunities to improve earnings.

Commented a forest products company spokesman in the annual report:

Liquidity of a company can be measured by several factors. Most important are:
   Earnings and cash flow generating capability
   Financial structure
   Working capital
   Credit lines
   Current value and liquidity of assets
(Potlatch Corp., 1980 Annual Report, p. 28).

The ability to generate a sufficient amount of internal funds for extended periods of time is a must for survival. Commented the proprietor of a southern lumber operation: "Some folks like to spend a lot of money. When they make money, they spend it on fancy equipment for the mill, golf courses, houses, cars, boats, and other things. I buy timberlands."

This operator currently holds 30,000 acres of prime pine timber, which have been acquired in up cycles over the past two decades. He had just completed another 2,000-acre purchase. He uses his timberlands as leverage; the timber feeds his mill in the downturns and smooths out his log purchases overall.

This operator also maintains a conservative balance sheet, which allows for management flexibility over the long term. "If I really need it, I buy it," commented the operator. "I'm set up so I am always able to buy what I really need."

It can be difficult to determine how liquid a company is just from a balance sheet. Analytical tools, such as the current ratio or the acid-test ratio, are of limited value. The true value of lands and timber, the condition of the plants and equipment, the ability to market a product in difficult times, and the resourcefulness of the management and employees are some other considerations. Commented an industry analyst after reviewing individual timberland holdings, "Our analysis indicated that timber is a significantly undervalued asset relative to . . . stated book values" (Clephane and Carroll 1981, p. 50).

While the financial fundamentals are important, so too are the operating basics. The successful companies manage the basics well at the conversion units, such as plywood plants and lumber operations. The basics are also managed closely at the cutting, bucking and log allocation steps. The basics are managed at each point at which a value decision is made; this includes managing each step in the process, from growing the tree to delivering the finished product.

Cost-Effective Conversion Facilities

Economy of scale, cost effective, capital intensive, state-of-the-art—all are management buzz words used to describe the efficient operator and his operation.

Commented a highly successful mill operator: "Our competitive situation cannot be materially improved by analyzing the market in search for new products or product combinations. Therefore, our efficiency to buy public timber must depend on our conversion ability to create high yields" (Northcross 1981, p. 11).

This competitive situation also applies to the fee timberlands–based operator. As a rule of thumb, the more efficient the conversion process, the higher the return on the timberland investment. Competition for logs, labor, energy and the marketplace demands the efficient use of the resource.

Plywood manufacture is an example of the trend to the more efficient. The average 1950 vintage plant produced about 40 million ft (3/8th basis) per year with 175 hourly employees. About 45% of the prime old-growth logs peeled eventually ended up as in-panel wood. The remainder was chipped, burned or disposed of from the plant site. This contrasts sharply with a mill constructed a quarter century later.

The newer plant is four times the size with fewer than twice the workers and resulting man-hours. Yield from the log may be up 40% and more from 1950, if rec-

Table 3.2. U.S. Softwood Plywood Production, by Region (MM3/8ths basis)

| Year | Region | | | | Total |
	West	Inland	Southeast	Other	
1930	305				305
1940	1,200				1,200
1950	2,554				2,554
1960	7,739	75		2	7,816
1970	10,120	891	3,315	14	14,340
1980	7,831	1,088	7,393	156	16,468

Source: American Plywood Assn. 1981

ords are available. An operator with a 1950-vintage mill will find competition difficult or impossible, even in a good market.

Current industry trends are best illustrated by statistics. Table 3.2 indicates total softwood plywood production in the USA at the beginning of each decade since 1930. The volume increased in multiples through 1970; it then leveled off but grew moderately.

Table 3.3 shows the number of mills operating at the beginning of each time period. The trend indicates more volume from fewer mills. The latest 10-year period, 1970 to 1980, indicates that the older West Coast mills are being rapidly replaced by larger, newer mills in the Southeast. This trend will be even more accentuated in the 1980s as the first-generation southern pine mills are modernized, expanded or abandoned.

The two largest North American structural panel producers, Georgia-Pacific and Champion Intl., reflect the trend. Their average production per plant in 1980 (Table 3.4) at 143.7 million ft (3/8th basis) per conversion unit was much higher than the industry average in the West of 77.5 or the Southeast at 113.7 (Table 3.5).

Sawmills, board plants and other conversion options are also following this trend. Each newer plant is larger and more cost-efficient than the earlier genera-

Table 3.3. Operating U.S. Softwood Plywood Plants, by Region

| Year | Region | | | | Total |
	West	Inland	Southeast	Other	
1930	17				17
1940	25				25
1950	68				68
1960	145	6		1	152
1970	127	11	40	1	179
1980	101	10	65	4	180

Source: American Plywood Assn. 1981

Table 3.4. North American Structural Panel Production, 1980: Two Companies, Total and Mill Average

Company	Mill count	Annual volume (MM3/8ths basis)	Average per mill (MM3/8ths basis)
Georgia-Pacific	21	3,084	146.9
Champion Intl.	17	2,366	139.2
Total	38	5,461	143.7

Sources: Georgia-Pacific Corp., 1980 Annual Report; Champion Intl. Corp., 1980 Annual Report

Note: Includes total volume of softwood plywood, waferboard and composite panels. Weldwood of Canada, a 73% owned Canadian subsidiary, is included with Champion Intl. Does include effects of labor stoppages in Canada and the U.S. Southeast during the cited year.

tion. The Simpson Mill 5, constructed in 1978 at Shelton, Washington, is an example. The mill is built to operate three shifts per day on second-growth logs with a butt diameter of 10 inches or less. This computerized mill will cut about 100,000–110,000 bd ft per shift out of about 1,800 stems. This high-volume mill was designed for the tree; the second-growth tree profile was not suitable for processing in an older mill.

Timberlands Ownership

The major ingredient for the forest products conversion unit is wood: the log or the residual from that log. Commented an industry analyst: "The only companies that have grown at faster-than-average rates over time are those with a dominate low-cost timber base" (Clephane 1978, p. 31).

Wood is the largest single cost ingredient for lumber, plywood and veneer. Other products have significant wood costs also. Timberlands ownership gives the producer a competitive edge in the marketplace; as such, it is an excellent marketing tool. It is also a financial reserve to ensure liquidity during market downturns.

Table 3.5. U.S. Structural Panel Production: Average Annual Capacity per Mill (MM 3/8ths basis)

Year	Region	
	West	Southeast
1930	17.9	
1940	48.0	
1950	37.6	
1960	53.4	
1970	79.7	82.9
1980	77.5	113.7

Source: American Plywood Assn. 1981

Table 3.6. North American Timberland Ownership: 20 U.S. Forest Products Companies

Company	Acres owned in fee (thousands of acres)		Estimated present value of land and timber (millions of $)
	1969	1979	
International Paper	6,519	8,410	3,857
Weyerhaeuser	5,444	5,923	9,534
Georgia-Pacific	3,500	4,500	3,404
St. Regis	2,408	3,179	1,571
Champion Intl.	1,687	3,066	2,120
Great Northern Nekoosa	2,444	2,712	614
Boise Cascade	1,981	2,664	1,703
Scott Paper	1,772	2,914	1,222
Crown Zellerbach	1,375	2,053	2,910
Union Camp	1,468	1,722	940
Time (Temple-Eastex)	NA	1,530	958
Continental Group	1,413	1,472	844
Diamond Intl.	1,330	1,453	681
Mead	1,091	1,352	669
Potlatch	1,366	1,311	1,210
Westvaco	1,160	1,221	522
ITT (Rayonier)	1,085	1,274	1,575
Owens Illinois	900	1,001	604
Louisiana-Pacific	NA	880	1,319
Kimberly Clark	1,500	561	439

Sources: O'Laughlin and Ellefson, 1981, p. 59; Clephane and Carroll, 1981, p. 49

"The mix of fee timber versus other sources of logs can significantly effect the cost of sales" cited one report (Pope & Talbot Inc., 1980 Annual Report, p. 10).

Table 3.6 identifies the ownership trend for a recent 10-year period for 20 large forest products manufacturers. The largest timberland owner, International Paper, owned in excess of 8 million acres in 1979. Weyerhaeuser, known as The Tree-Growing Company, had a lesser acreage at 5.9 million but a significantly higher value for its land and timber. This reflects the higher value per acre of Weyerhaeuser's extensive West Coast holdings.

Except for Kimberly Clark and Potlatch, each company increased land ownership significantly during the period. Potlatch subsequently purchased 93,000 acres of Idaho timberland from Milwaukee Land Co. during 1980. "Without question," stated L. C. (Whitey) Heist, an industry leader, "top management is more aware of the forest land base as a corporate asset than ever before" (Northcross 1979, p. 11).

It then comes as no surprise that the top five industry sales leaders, Georgia-Pacific Corp., International Paper Co., Weyerhaeuser Co., Champion Intl. Corp., and Crown Zellerbach Corp., each with sales in excess of $3 billion, are also major timberland owners.

Commented an industry analyst: "The top 40 U.S. wood-based companies in 1978 accounted for 40 percent of all U.S. wood-based sales revenue. These companies own 53.6 million acres or 80 percent of all the U.S. commercial timberlands. This represents 11 percent of the entire U.S. commercial forest land base. ... those companies with the largest timberlands holding dominate the sales ranking" (O'Laughlin and Ellefson 1981, p. 61).

Well-Defined Core Business

Each successful forest products manufacturer has a well-established core business. It may be a specific grade or grades of paper for some; for another, a lumber type; and another may focus on plywood. The company then builds and expands from the core business. Lumber to paper or paper to lumber has been the traditional route. Plywood and board manufacturing then follow in sequence.

Weyerhaeuser's history is a variation on that route to integration. Founded as a timber seller, the company progressively moved to lumber, paper and the other products in sequence. It maintains an extensive log and timber-selling business along with selling chips in the Pacific Rim markets. Its core business, however, remains tree growing. Log conversion is considered a logical extension of that core business.

Integration and Diversification

As stated earlier, forest products firms have integrated and diversified as a logical extension of the core business. Cited reasons are as follows:

- To smooth out the earnings flow by balancing countercyclical markets, such as paper versus building products.
- To dampen the cyclical effect by moving into products other than commodity products.
- To extract the maximum value from the timber resource by operating as a total converter.
- To optimize and maximize the earnings flow by widening the range of product options.

The sum total is a more efficient use of the resource and a consistent, relatively high earnings flow. But integration and diversification are not just for the industry giants; the small operator can also benefit.

A small producer, unable to own additional conversion facilities, can develop a ready market for the residuals or the portion of the log he is unable to use efficiently. For example, chips now provide essential operating revenue for many and thus have come to be viewed in the industry as a regular mill product rather than a residual.

This is an evolving trend; as late as 1970 most of the smaller sawmills in Washington State were without barkers and chippers, equipment essential to making clean chips from slabs and edgings. Not surprisingly, "Many just went in and out of existence" (Bergvall and Holtcamp 1970, p. 7). Many operators have survived through recent cyclic market downturns because of the income provided from chips, shavings and other residual products.

Marketing Organization

The highly successful firms have developed marketing and distribution organizations. Most have developed organizations that can market and distribute more than the company produces.

The strategy is best explained by the company biographer for Georgia-Pacific: "Hunt made it a policy for the Distribution Division to sell 120 percent of G-P production so that they would keep G-P mills running when markets were weak" (Ross 1980, p. 97).

This is a widely used policy. It evolved over time as each of the sales organizations, such as U.S. Plywood, integrated into manufacturing. Some companies distribute to wholesalers, others wholesale to the retailer and some have retail outlets. One company, Evans Products, has evolved heavily into retailing while reducing its manufacturing capability.

Markets and Market Shares

Many observers view the forest products industry as a business dominated by a powerful few firms, but the industry, particularly lumber manufacturing, remains fragmented and is highly price-competitive when compared with other industries. Forest product markets remain sensitive to commodity auction pricing and the demand/volume available in the marketplace.

For example, 1980 U.S. lumber production was 27.3 billion bd ft (Lambert 1981, p. 5). Canada produced 15.3 billion, for a total North American production of 42.6 billion. The top 100 producers for the same period produced 29.1 billion, or about 68% of the total. The largest producer, Weyerhaeuser, produced only about 6% of the total volume.

The plywood industry is somewhat similar, although fewer firms control a greater share of the market. This has become particularly true as the industry continues its shift to the Southeast. The top two firms, Georgia-Pacific and Champion Intl., produce about a quarter of the total U.S. and Canadian volume.

Some firms dominate a much narrower market segment. Two plywood plants produce the majority of the western red cedar siding; two others produce the lion's share of the medium and high-density overlays. In lumber, a similar situation has developed with export clears and redwood lumber. Although the goal of the producer is to gain a significant market share to allow for economies in manufacturing and marketing, the chance of dominating an end-use market is remote. Too many substitute products can fill the consumer's need.

The eight characteristics discussed here are common to the successful forest products producers, but little has been said of people and their role in the organization. People selection and development, along with an organization that makes use of their individual and collective talents, is the overriding characteristic of the successful forest products company. Good people get good results in an innovative, results-oriented organization. That's an industry heritage.

REFERENCES

Adams, D. M., and others. 1979. *Production, Consumption, and Prices of Softwood Products in North America: Regional Time Series Data 1950–1976.* Research Bulletin 27. Corvallis, Oreg.: Oregon State University, Forest Research Laboratory.

American Plywood Assn. 1981. *Annual Softwood Plywood Production Statistics.* Tacoma, Wash.

Amigo, E., and Neuffer, M. 1980. *Beyond the Adirondacks: The Story of St. Regis Paper Co.* Westport, Conn.: Greenwood Press.

Bergvall, J. A., and Holtcamp, R. J. 1970. *Washington Mill Survey, Wood Consumption and Mill Characteristics.* Report No. 2. Olympia, Wash.: State of Washington, Department of Natural Resources.

Clephane, T. P. 1978. Ownership of Timber: A Critical Component in Industrial Success. *Forest Industries* 105(9):30 (August).

Clephane, T. P., and Carroll, J. 1981. Timber: Even Non-Industry Investors Realize Its Value. *Forest Industries* 108(1):49 (January).

Coman, E. T., and Gibbs, H. M. 1978. *Time, Tide and Timber: A Century of Pope & Talbot.* Palo Alto, Calif.: Stanford University Press.

Drucker, P. F. 1980. *Managing in Turbulent Times.* New York: Harper & Row.

Helfert, E. A. 1973. *Techniques of Financial Analysis.* Homewood, Ill.: Dow Jones–Irwin.

Lambert, H. 1981. Three in a Row . . . , Then Down We Go. Annual Lumber Review & Buyers' Guide. *Forest Industries* 108(6):4 (May 30).

Logger and Lumberman, The. 1981. Simpson's Stud Mill Starts Third Shift. Vol. 30, p. 11.

1980 Annual Reports: Boise Cascade Corp., Champion Intl. Corp., Crown Zellerbach Corp., Evans Products Co., Georgia-Pacific Corp., International Paper Co., Louisiana-Pacific Corp., Pope & Talbot Inc., Potlatch Corp., St. Regis Paper Co., Union Camp Corp., Weyerhaeuser Co., Willamette Industries.

Northcross, S. 1979. Loggin' and Living: A Safe Bet on Southern Roulette. *Timber Harvesting* 27(6):11 (June).

———. 1981. Sun Studs Unleashes High Yield Studmill. *Timber Processing* 6(1):11.

O'Laughlin, J., and Ellefson, P. 1981. U.S. Wood-Based Industry Structure. Part 1: Top 40 Companies. *Forest Products Journal* 31(10):55 (October).

Pease, D. A. 1971. 1970 Lumber Production. 1971 Buyers' Guide & Yearbook. *Forest Industries* 98(7):10 (May 28).

Ross, J. R. 1980. *Maverick: The Story of Georgia-Pacific.* Portland, Oreg.: Georgia-Pacific Corp.

Western Wood Products Assn. 1981. *1980 Statistical Yearbook of the Western Lumber Industry.* Portland, Oreg.

Weyerhaeuser, G. H. 1981. Forest for the Future: The Weyerhaeuser Story. An address delivered at a national meeting of the Newcomen Society of North America, New York, N.Y., February 12, 1981.

Section Three
MANAGEMENT PLANNING

The most successful firms have a clearly defined mission.
This mission—or overall conceptual justification for operating
the business—provides the underpinning for the strategic
decisions that follow.

CHAPTER FOUR

Four
Strategic Planning

Strategic planning is a formidable term. The term just doesn't seem to fit a business discussion related to logging, lumbering or panel manufacturing. *Strategic* conjures up visions of military operations or a football game. *Planning* describes what a manager intends to do, but ... "I just never seem to get around to it."

Strategic planning, simply stated, is nothing more than an organized process that determines the most profitable course for the business over time. Whatever the terminology used, an industry observer can differentiate between the firms that do it well and those that do not. Those that do it well are characterized by purposeful growth and increased profitability.

A successful forest products firm usually relies on a formal strategic planning process. The plan is formulated and then implemented by key managers under the direction of top management. This chapter will focus on the successful company. Its planning process will be examined, along with the role its personnel play in determining the course of action.

THE MISSION: THE UNDERPINNING OF STRATEGIC PLANNING

The most successful firms have a clearly defined mission. This mission—or overall conceptual justification for operating the business—provides the underpinning for the strategic decisions that follow. These firms have an understanding of who they are and the purpose of the business. Weyerhaeuser Co. is an example of one such firm.

"We are engaged in the production of an organic material—wood—from our six million acres of soils, with aid from the sun, the rain, and the air. We assist this natural process by planting genetically selected and carefully propagated superior trees in soil conditions prepared for them ... We grow trees for a purpose, or rather for a multitude of purposes and products" (Weyerhaeuser 1981, p. 7).

Stated another company: "St. Regis is an integrated forest-based products enterprise, international in scope, which manufactures kraft and recycled products, printing papers, newsprint, construction products, finished packaging, consumer and converted products. The company's operations include 143 plants in nine countries, and its affiliates operate 89 plants in 23 countries. St. Regis' natural resource consists of 5.97 million acres of timberland owned or controlled and important interests in energy reserves" (St. Regis Paper Co., 1980 Annual Report, p. 2).

These statements, one general and the other more specific, represent a broad overview of the business; each is representative of the larger firms. As a rule of thumb, the smaller the business firm, the more specific the mission.

"Mission? I make dimension lumber and lots of pulp chips. I buy logs by the ton from small woodlot owners and pay cash. I can do things the big boys can't do!" This is the rejoinder given by a small southern pine lumber manufacturer during an interview.

The successful firm, large or small, will have a clearly defined mission. That mission may be resource based, such as Weyerhaeuser's; manufacturing based, such as St. Regis's; or sales based, such as Evans Products Co.'s. In fact, each distinct business within a large corporate unit may have a different mission or purpose.

The mission must recognize that the end result must be a profit; the purpose of the forest products enterprise is to create and serve the forest products customer, who in turn will consistently pay a product price that will be sufficient to cover the producer's expenses plus a fair return. The conceptual framework within which the forest products firm strives to build this profitable existence is the mission.

KEY ISSUES IN STRATEGIC PLANNING

The mission of the firm is shaped by the business environment; the business environment is a product of the key issues facing the producer and how each issue is addressed. These key issues are evaluated as part of the strategic planning process as follows.

Expectations of the Equity Holders and the Public

Owners, shareholders, employees, suppliers, customers and creditors each expect something different from the ongoing business. The public and the government also have differing expectations. Each expectation is addressed as part of the company's mission.

The equity holder expects a return on investment that is sufficient to encourage continued ownership; the public may expect jobs and the economic benefits that follow. Simpson Timber Co., a large privately held firm, is a producer that must meet certain clear-cut public expectations.

This company, holder of a 100-year Cooperative Sustained Yield Unit contract with the U.S. Forest Service, is required to maintain production facilities (and the associated employment) in Mason County and the surrounding counties in western Washington. The company in turn has assured access to government timber for the 100-year period.

The Simpson situation is unique; few others must focus so closely on public needs. Many companies profile their stockholders and determine what they expect.

"An understanding of how the great majority of our shares are held indicates that the company should pursue a direction which emphasizes long-term capital appreciation within reasonable limits of risk," is the planning criterion for one firm.

Defining expectations provides a framework for future planning.

Definition of Growth

Each company defines growth in a different fashion. For some, growth is the natural evolution of the business as it matures. For others, growth is a business strategy.

"The prospects for future growth in all our businesses remain excellent. We continued to invest in this future during the year, with new facilities, expansions, and equipment upgrades. . . . We have positioned our company well for future expansion" (Georgia-Pacific Corp., 1980 Annual Report, pp. 6–7). In 1980, Georgia-Pacific was designated by *Fortune* magazine as the firm having the best overall growth record of all Fortune 500 companies over a 25-year period. Four decades earlier Georgia-Pacific was a medium-size regional lumber producer.

Another company, founded prior to the turn of the century, remains a relatively small, private firm. The company's former chief executive officer commented in an interview as he evaluated his company against others: "We were bigger, better, and stronger than any of the component companies [of a competitor]. They became a public company and perhaps followed a more profitable path than we did, although I think our way is better for a family like ours" (Reed and Maunder 1977, p. 153).

The issue of growth and its role in the company's future must be determined. Some questions to be asked:

- Do we want to grow? At what rate? In what businesses?
- How is growth to be achieved? Internally or through mergers? Integration (horizontal and/or vertical), diversification or other ways?
- What product lines should we pursue? Lumber, paper, plywood, board products and/or other products? Should the products be commodities or specialized products? Who is the customer: converter, wholesaler or consumer?
- How is growth to be financed? Internally or through equity financing, short-term/long-term financing and/or other methods?
- What is the rationale for growth? Is it to achieve a more efficient utilization of the wood supply? Do we diversify to reduce or eliminate the roller coaster ups and downs of one or two market-sensitive product lines? Is it higher return for the stockholder?

Most companies find that at least some growth is necessary to maintain stockholder interest. Forest products companies have determined that successful growth attracts additional resources, which provide the foundation for further growth. In addition, a manager recently stated, "We feel that some growth is necessary to attract and retain the type of people we need to lead this company!"

Evaluation of Company Strengths

What are the company's strengths and what does it do well? Currently most companies expand into profitable businesses, businesses in which the firm has an inherited, existing or potential expertise. Individual companies are tending to focus on a handful of product types; additional products are an extension of the favored product line and may be pruned as the parent business perceives each to be unprofitable, capital intensive or a poor fit. In recent years, paper-oriented companies have divested solid-wood-product mills; lumber-oriented companies have backed away from the structural panel business. "We do well at what we know and understand best" is the guiding theme.

Recognition of Resources Available

What is the extent of our resources: land, facilities (manufacturing and distribution), finances and people? How can we obtain maximum return from these resources individually and in combination? A Weyerhaeuser will be land oriented, using financial, human and facility resources to maximize the return from its land. A Louisiana-Pacific will rely on manufacturing facilities and the entrepreneurial ability of its leaders to maximize the return to its stockholders. Evaluation of the company's strength is a necessary step in planning.

Identification of a Planning Horizon

"Long term planning is essential to success because in a mature, capital-intensive business like forest products, planning must cover 10, 20, 30, and up to 60 years and span many economic cycles" (Crown Zellerbach Corp., 1980 Annual Report, p. 5).

A long-term plan—one based on a planning cycle of 30 or more years—is a must for a renewable resource-based business. In the forest products industry, the focus of the long-term plan is detailed resource planning, for both people and trees. Long-term planning is basic to, yet separate from, the detailed strategic planning effort, which generally covers a 5 to 10-year period. The specifics of the strategic plan will be most accurate at year 1, of reasonable value at year 5, and at year 10 and longer, hazy on specifics but still of significant worth.

"In other words, up to some cutoff point uncertainty can be handled by multiple estimates of gains and costs in each period, but beyond that point the effect of uncertainty so swamps the estimates of expected gains and losses that they may as well be discounted to zero," points out one writer (Hall 1968, p. 273) in describing the present value of forward detailed planning.

Most forest products companies will recognize the growth cycle of the forest lands from which they draw their raw material. Broad planning will take the longer cycle of decades into account; detail planning will seldom project more than 5 years ahead.

BARRIERS TO PLANNING

Tradition and uncertainty are the two greatest barriers to effective planning. A speaker stated at a wood products clinic in 1964:

1. The future won't be like the past.
2. It won't be like you think it will be.

3. And the rate of change will be faster than ever before (Wellwood 1972, p. 15).

Change is occurring at an ever rapid pace, as the years since that prediction have demonstrated. "Chances are that if a manager or a business leader is doing the same thing the same way he or she did five years ago, the individual manager has become obsolete" is a comment that describes the current operating and planning environment.

"We have always done it this way" is a typical comment that reflects the first barrier, tradition. One way to overcome resistance to change is to recognize that change itself is an unavoidable tradition. Frequently, resistance to change can be overcome by studying the successes of others; participative activities that lead to change (such as quality circles or involvement groups) also reduce resistance in others.

The second barrier, uncertainty, is represented by statements like this one: "Why should we plan, not knowing what the future will bring?" This barrier is overcome in a variety of ways. A simple projection of present trends is a method of some value, but it has become less useful as the business environment and the society in which it functions become increasingly interrelated and complex.

Econometric modelers such as the Townsend-Greenspan, Data Resource and Chase Econometrics are used by some. More specific forecasts of forest industry trends have been developed by trade associations, government economists, financial analysts and other individuals who are closely associated with the industry and its customers.

Raw material and product price trends, consumer demand patterns, timber availability forecasts, consumer profiles and definitions of related issues (such as transportation, regulatory and energy trends) are some information types available to the planner. There is an almost endless supply of topical information; the challenge is to select the vital few bits from the many.

Information of this sort provides a framework for planning; its weakness is the inordinate weight usually given to current events as an indicator of long-term trends. The planner will assemble the general and specific information and then correlate the various forecasts and projections into a model that represents the expected planning parameters. This model, a composite of available forecasts and projections, can then be tailored into a best fit model that is unique for the individual company.

THE STRATEGIC PLANNING PROCESS

The strategic planning process has four basic steps: (1) identify the basic business strategy, (2) outline the strategic plan objectives, (3) formulate a program to achieve those objectives and (4) document the results expected for each year of the planning period.

Table 4.1 is an example strategic plan summary. The table documents the results annually for a hypothetical waferboard operation. Annual production volumes are identified along with the unit costs. The net profit, the bottom line figure, is also shown. Not illustrated but noted are the additional year-by-year reference fig-

Table 4.1. A Strategic Plan Example Summary

Business: Structural panels, waferboard
Facility: Boundary City, Minn.

	Actual 1984	Est. 1985	1986	1987	1988	1989	1990
Annual production (MM3/8ths)	100	105	108	110	112	112	112
Unit sales return ($000/M3/8ths)	140.0	135.0	138.0	145.0	150.0	157.0	163.0
Unit costs ($000/M3/8ths)							
Material							
Wood	35.0	32.0	35.0	37.0	38.0	40.0	41.0
Glue	25.0	28.0	30.0	33.0	34.0	35.0	38.0
Wax and other	3.0	3.0	4.0	4.0	4.0	5.0	5.0
Total	63.0	63.0	69.0	74.0	76.0	80.0	84.0
Labor	10.0	10.6	11.1	11.5	12.0	13.0	13.5
Variable mfg. exp.	38.0	40.0	42.0	42.0	43.0	44.0	45.0
Fixed expenses (salaries, deprec. and others)	18.0	18.2	18.4	18.4	18.6	18.6	18.7
Total unit costs	129.0	131.8	140.5	145.9	149.6	155.6	161.2
Misc. income	3.7	3.8	3.9	3.8	3.9	4.0	4.0
Net profit	**14.7**	**7.0**	**1.4**	**2.9**	**4.3**	**5.4**	**5.8**

Additional year-by-year reference to:

• Plant ratios and statistics
• Investment and cash flow information
• Investment evaluation criteria and other facts or evaluation information would be included year by year as required.

ures that may be required to gain a further understanding of the manufacturing plant's prospects for each year of the planning period.

Step 1: Identify the Business Strategy

The following is an example of a basic business strategy:

Our corporate growth strategy is to:

1. Build a solid commodity business by improving and increasing utilization of our raw material base and delivering commodity products to market at the lowest unit cost.

2. Use this commodity business as a foundation for supporting investments in higher risk, higher return opportunities in present and expected markets, by deriving new methods and technologies and in new businesses.

In a strategic plan, the business strategy is usually preceded or followed by a qualifying statement that reinforces the mission of the company, outlines its strengths and describes how the business will react in its business environment.

Step 2: Outline Objectives

An overall objective is then developed to describe the expected end result of the strategy.

Specifically, the business overall objective is to sustain and accelerate an average growth rate in earnings per share of 14% per year over at least the next 5 years

is an example of an overall strategic plan objective.

Step 3: Formulate a Program

Specific objectives are then identified, along with related strategies to achieve those objectives.

Specific areas of opportunity, or leverage points, are identified as follows:

- Raw material utilization: (1) increased recovery available from the log; (2) increased conversion return from a specific raw material into a higher value product; (3) replacement of higher cost raw materials with lower value materials
- Capital improvements to achieve lowest cost producer status through cost improvement
- Opportunities such as nontraditional products and processes
- Realignment of the business to eliminate obsolete, marginal and unprofitable businesses, facilities and products
- Growth opportunities available in the more profitable businesses
- Technology improvements specifically related to: (1) products; (2) processes; (3) measurement; (4) controls

The company then organizes planning groups, which in turn identify the individual opportunities and the attainable results. System engineering—a methodology that identifies specific opportunities and then translates those opportunities into an operating plan—is frequently applied to the planning process at this point. The methodology:

- Defines the opportunities available
- Selects the attainable objectives
- Identifies facilities and/or processes that can achieve the specific objectives

- Analyzes the various options available and selects the optimum combination of opportunities through system and financial modeling
- Communicates the resulting plan using a formal descriptive document that states the objective, the process for achieving the objective, the forecasted results, and the assumptions used to determine the process and the forecasted results

The system engineering planning group is usually a specific task group drawn from the operating business. This task group is composed of a select number of core specialists, such as operating managers, system planners, engineers and financial managers. Other specialists, such as resource and process specialists, are drawn into the group as needed. The leader, usually a selected operating manager, provides direction and seeks the individual and combined expertise of the group. The system engineering methodology assumes that the company's management and the selected systems engineering task group have a strategic overview sufficient to permit wise choices between options.

The planning groups are organized by major businesses such as softwood lumber, structural panels, packaging or other stand-alone business divisions. These planning groups are then broken down to the smallest separable production or sales unit at each specific business location.

Step 4: Document Expected Results

The aggregate plan for each of the business entities is folded into the business unit plan; the plans for each of the business units are then combined into the overall company plan. The resulting master plan, when accepted by top management, usually covers a 5 to 10-year period. Year 1 of this strategic plan is the tactical or annual plan. This planning process is updated, modified and extended annually to reflect the input of the latest facts and trends.

THE STRATEGIC PLAN: LINCHPIN AND LINKAGE

The years ahead will witness slower and more erratic growth in both domestic and world markets. Inflationary and regulatory pressures will continue, as will critical public response to business and investment decisions. The successful company will plan and plan well as a means to cope with these business conditions.

Increasingly, the successful company will be either the low-cost producer or the company offering the most differentiated forest products. Detailed and timely planning will be required to achieve either or both. The strategic plan is the linchpin that holds together the overall business process. It is the linkage in a stream of activities that profitably converts a tree into forest products: forest products that will serve an ever-widening range of consumer demands.

REFERENCES

Bender, P. S., Northup, W. D., and Shapiro, J. F. 1980. Practical Modeling for Resource Management. *Harvard Business Review* 59(2):163.

Hall, A. D. 1968. *A Methodology for Systems Engineering.* Princeton, N.J.: Van Nostrand.

Hall, W. K. 1980. Survival Strategies in a Hostile Environment. *Harvard Business Review* 59(5):75.

1980 Annual Reports: Crown Zellerbach Corp.; Evans Products Co.; Georgia-Pacific Corp.; St. Regis Paper Co.

Reed, W. G., with Maunder, E. R. 1977. *Four Generations of Management: The Simpson-Reed Story.* Santa Cruz, Calif.: Forest History Society.

Sawyer, F. 1979. The Use of Strategic Models in Setting Goals. *Michigan State University Business Topics* 27(3):37.

Wellwood, R. W. 1972. *Changing Aspects of Wood Utilization.* WPS 72.2 P. Vancouver, B.C.: University of British Columbia.

Weyerhaeuser, G. H. 1981. Forest for the Future: The Weyerhaeuser Story. An address delivered at a national meeting of the Newcomen Society of North America, New York, N.Y., February 12, 1981.

Five
Financial Planning and Budgeting

Mill operators frequently call it the annual plan, the profit plan or the operating budget. It is sometimes called the tactical plan when prepared as year 1 of the strategic plan. Whatever it is called, the operating budget is the road map for the immediate present. It focuses the operation manager's attention on the specifics of running the business.

"The specialists who draw up the operating plans focus on the dimensions of their internal world; parts lists, drawings, work standards, shop rules, production schedules, plant layout, and so on. On the other hand, the authors of business plans look outward. Their focus is on the external environment in which the business operates. Thus the profit center manager must match these two perspectives; there is no one to whom he can delegate this responsibility" (Hobbs and Heany 1977, p. 122).

The operating budget is prepared each calendar year and is segmented into monthly or quarterly time periods. The manager's role is to take the various perspectives provided by the personnel in each functional area and then weave that information into a realistic and workable plan.

When accurately prepared and coupled to detailed demand projections and the resulting market prices, the operating budget becomes the yardstick for performance evaluation, a tool for financial tracking and a means of coping with uncertainties in the industry.

PURPOSES OF THE OPERATING BUDGET

Accountability is the name of the game in planning. Participation by both line and staff personnel ensures a well-prepared budget. Participants are then held accountable for their individual contributions as the year unfolds.

Commented a lumber operator during an interview, "I had a procurement forester who forecasted the price and availability of peeler-type logs for our export cutting program; he was then to log the mill like he said he could. He didn't do it and I didn't make my plan. He is no longer with us."

When administered wisely, the operating budget:

1. Compels accurate management planning and preparation
2. Provides definite expectations that are the framework for judging subsequent performance
3. Promotes communication and coordination among and between the various line and staff positions within the location, the division or the company

"The planned manufacturing cost projections were about a half million dollars low because the accountant had an error in his extensions. He will know better the next time . . . if there is a next time," was another comment heard during an interview. This manager knew that he was expected to make his operating budget—despite the error. The company's cash flow projections and subsequent financing needs had been based in part on his plan. This manager's comment also recognizes the second purpose of the operating budget, a tool for tracking performance and financial results.

Commented one company, "Evans prepares an annual Financial Plan as part of a five-year planning cycle . . . all locations submit income statement projections for each month of the ensuing year. These projections are developed from a detailed budgeting process involving analysis of market conditions, product lines, costs and asset requirements" (Evans Products Co. 1982, p. 8).

The individual statements, and the supporting data, then become targets that encompass all phases of the operations. The operating budget is used to gauge results, and it is used as a yardstick to evaluate progress toward the company's planned objectives.

"Balance sheets are submitted along with income statements as part of the Financial Plan process. The actual levels of inventories, receivables and payables are monitored monthly and compared with established targets. Variances are investigated and corrective action is taken promptly" (Evans Products Co. 1982, p. 8).

The manager is in the spotlight; also in the spotlight are the corporate and division staff members who develop the relevant economic indicators and other guidelines that are supplied as a foundation for the operating budget.

Coping with uncertainty is another purpose of the operating budget. Traditionally, managers have eschewed the formal planning process in favor of a seat-of-the-pants approach. "How would I know what we are going to make out of that raft of logs? Let's see how the grade develops," is a comment typical of the seat-of-the-pants adherent. Cheap logs, plentiful labor and a good market tolerated that management style in the past.

When compelled to submit an operating budget, the adherent would turn the whole process over to the accountant or to the newest production trainee, the latter usually fresh from college. It is no coincidence that most seat-of-the-pants adherents are no longer in active management, and each market downturn results in fewer and fewer survivors. The entrepreneur goes broke; the corporate manager is squeezed out when things go wrong.

Managers must grapple with uncertainties, with or without a plan. Detailed budgeting gives a focus to the grappling. The benefits of detailed planning exceed the investment in time and resources.

THE OPERATING BUDGET AND THE
ORGANIZATIONAL STRUCTURE

The operating budget defines the short-term business tasks; it is linked closely to the organizational structure. Most forest products companies are segmented into manageable parts for planning and accountability. This usually takes the form of functional organizations; one person is responsible for sales, another for production and a third for the log supply and land management function. The organization may be further segmented into staff services such as purchasing, employee relations and other functions.

Each definable business segment that represents a profit center, a cost center or a functional staff group develops an individual plan. The individual and local plans are then folded into a total business-unit plan; these in turn are combined into the regional and division plan. The plans are further consolidated at each level of direct responsibility until the total company operating budget represents the expectations of the chief executive officer.

Separating the planning structure into logical and definable stand-alone segments aids in managing the operation and the company. It permits detailed analysis and sharply defines accountability. Projections of operating results are expressed in a format that identifies the scope of the business segment involved. The operating budget must define the elements under the control of the responsible manager and becomes the basis for performance measurement. It is structured to fit the particular operation unit and operating style of the company.

THE LOCAL-UNIT PLAN: PROFIT CENTER OR COST
CENTER CONCEPT?

The operating budget is a quantitative expression of a course of action. As such, the operating budget for a local manufacturing unit may be prepared in one of two forms. The profit center concept is usually used within small to medium-size firms. The large fee-sourced giants may use either the profit center concept or the cost center concept.

The *profit center concept* defines a local manufacturing unit as a stand-alone business with some allocated division and corporate costs. Firms using the *cost center concept* identify the relevant production objectives and set up the reporting system based on expected costs. The manager is then held accountable for achieving the objectives within the budgeted costs.

Profit center or cost center: which is the most effective performance accounting tool? The answer depends on the objectives of the organization. A centralized organization that exercises coordinated control over two or more manufacturing units and controls the log input and the product mix will probably choose the cost center concept. The decentralized firm that operates as a series of relatively independent businesses will choose the profit center concept. Some firms will operate as a mix of the two.

The most aggressive growth firms prefer the profit center concept because of the visible accountability it provides. In addition to visibility, the profit-and-loss statement numbers plus supporting data point up cause-and-effect relationships.

Planning and budgeting define the specifics of running an individual cost center.

These documents provide insight and a high degree of self-discipline when a decision must be made.

"That's what the job is all about," commented a young manager of a large West Coast complex. "That bottom line is the scoreboard which tells how I'm doing." This manager recognizes that lumber and panel manufacturing is a value-adding process.

Running a manufacturing facility as a cost center can produce some unusual occurrences, as these comments illustrate: "You wouldn't believe it if I told you," mentioned a plywood manager as he replaced the telephone on its cradle after talking with his boss. "Just a few days ago he called and got all over me for using too many man-hours. I explained that I was taking advantage of the sanded market uptick and had switched from sheathing. This strategy required thinner peels, increased veneer patching, and I added more panel patchers. . . . Boy, was he mad! But today he called for a different reason. . . . His boss said the division made record profits and my mill was the star contributor."

This manager's cost statement is documented history. The compliment was nothing more than passing glory. The following month he switched back to sheathing to get his costs in line. The cost center concept does little to encourage product mix decisions of this sort. And the manager may find himself being judged under two sets of rules—as a cost center by his immediate boss, but as a profit center by senior management.

Another side of the same coin—raw material utilization—is reflected in these comments: "Save fishtails? Why would I want to do that?" asked an East Texas operator. "Fishtails cost too much to process. . . . My yield is on plan." A variation on the same theme in a lumber operation: "I can get more through the trimmer if I chip the green boards; it just helps my costs."

These and other like decisions are made daily, and they result in substantial added profit or losses. With the cost center concept, the costs of these decisions do not appear because the lower yield has been accepted as a standard. The cost center concept does little to encourage the manager to seek out profit opportunities, while the profit center concept appears to support such decisions more effectively. But the profit center concept is not without pitfalls.

At one extreme, defining the profit center too broadly can result in hazy decisions. It can also result in decisions that produce an apparent benefit (improving the numbers at the local operation, for example) but that may be less than optimum for the complex, the region or the division as a whole. At the other extreme, defining the profit center too narrowly can cause other problems. Creating a profit center for each of the products produced from the same log (such as lumber, chips, hog fuel and other products) will require extensive cost allocations and will result in profit centers that have little meaning as such.

A profit center should be a relatively stand-alone operation that could function as a business in an integrated business community. A profit center must assign realistic costs or transfer prices to the raw material purchased and the product sold.

For example, a West Coast lay-up plant was required to purchase all its incoming veneer from a company-owned veneer plant. The quality of the veneer was below that of the industry in general and with a higher proportion of D grade and random widths, yet the lay-up plant paid the market price for CD veneer. Subsequently, the lay-up plant's product mix faltered, as did the yield at the veneer plant. The regional manager, who lacked hands-on operating experience, concluded that more veneer plants and fewer lay-up operations were needed—and that the veneer business was good and the panel business bad. He soon found himself with lots of green veneer that had to be heavily discounted in a competitive marketplace.

ELEMENTS OF THE OPERATING BUDGET

As a quantitative expression of intended action, the operating budget aids in coordination and control. It has a number of elements (Figure 5.1). The first is the assumptions.

Assumptions

Assumptions include the *mode of operation*; this identifies the log type to be used and in what quantity for a given operating schedule. Also included are *forecasted costs* and *sales returns*, and *efficiency factors* such as grade yield, volume recovery and labor man-hour efficiency. Other assumptions, including the projected effect of the *capital spending plan* on the mill operation, are documented as appropriate.

The mode of operation is determined in a number of ways. For some, it may be a repeat of a previously successful mode of operation. For others, it may be determined by utilizing a linear program or an operation simulation model (see Chapter 6), a pro forma profit and loss statement or other decision analysis tools. The mode

Assumptions
Mode of operation
Forecasted costs and sales
Efficiency factors
Capital spending plan
Other appropriate details

Pro forma profit (loss) statement
Statement by month
Annual consolidated statement

Financial exhibits and trends
Performance trends
Prior year's comparison
Financial analysis exhibits

Figure 5.1. Elements of the operating budget.

Figure 5.2. By-products as a log cost offset.

Annual Profit and Loss Summary
Intermountain Plywood

	$000	$/M3/8ths
Plywood sales	$18,897	$224.17
Less adjustments	(20)	(.24)
Net revenue	18,877	223.93
Cost of goods sold		
Wood	9,258	109.83
Less by-product	*(1,829)*	*(21.70)*
Net wood cost	7,429	88.13
Glue	686	8.14
Mfg. supplies	258	3.06
Direct labor	4,248	50.39
Mfg. expense – fixed	488	5.79
– variable	2,385	28.29
Total cost of goods sold	15,494	183.80
Gross profit (loss)	**3,383**	**40.13**
Shipping and selling	82	.97
Factory overhead	634	7.52
Total operating expense	716	8.49
Operating profit (loss) before taxes	**$ 2,667**	**$ 31.64**
Plant production (M3/8ths)	84,298	

of operation will usually follow the guidelines previously developed in the strategic plan.

The operating budget begins to take shape as the planning assumptions are documented and agreed upon.

Pro Forma Profit-and-Loss Statement

The next element is the pro forma profit and loss preparation. To prepare this statement the departments or cost centers are first identified, and the total and per-unit costs are calculated for each. The informaton for all of the departments is then consolidated into a stand-alone profit or loss summary sheet. This summary is usually prepared in one of three formats. The formats differ in the way they identify revenues and expenses.

Method 1: By-Products As Raw Material Cost Offset. Figure 5.2 illustrates Method 1 using a softwood plywood plant as an example. The method can be used whenever a marketable by-product develops from the primary manufacturing process in the plant.

Figure 5.3. By-products as below-line income.

Annual Profit and Loss Summary Intermountain Plywood		
	$000	**$/M3/8ths**
Plywood sales	$18,897	$224.17
Less adjustments	(20)	(.24)
Net revenue	18,877	223.93
Cost of goods sold		
Wood	9,258	109.83
Glue	686	8.14
Mfg. supplies	258	3.06
Direct labor	4,248	50.39
Mfg. expense – fixed	488	5.79
– variable	2,385	28.29
Total cost of goods sold	17,323	205.50
Gross profit (loss)	**1,554**	**18.43**
Shipping and selling	82	.97
Factory overhead	634	7.52
Total operating expense	716	8.49
Operating profit (loss)	838	9.94
By-product income	*1,829*	*21.70*
Operating profit (loss) before taxes	**$ 2,667**	**$ 31.64**
Plant production (M3/8ths)	84,298	

Annual Profit and Loss Summary
Intermountain Plywood

	$000	$/M3/8ths
Plywood sales	$18,897	$224.17
Less adjustments	(20)	(.24)
Chips and by-products	*1,829*	*21.70*
Net revenue	20,706	245.63
Cost of goods sold		
Wood	9,258	109.83
Glue	686	8.14
Mfg. supplies	258	3.06
Direct labor	4,248	50.39
Mfg. expense – fixed	488	5.79
– variable	2,385	28.29
Total cost of goods sold	17,323	205.50
Gross profit (loss)	**3,383**	**40.13**
Shipping and selling	82	.97
Factory overhead	634	7.52
Total operating expense	716	8.49
Operating profit (loss) before taxes	**$ 2,667**	**$ 31.64**
Plant production (M3/8ths)	84,298	

Figure 5.4. By-products as sales revenue.

Sales are defined as income from the primary product less any claims and customer adjustments. Net wood costs are the primary raw material cost (FOB mill-yard) less by-product income.

This method is common among firms that purchase a significant amount of logs on the open market. Timber-sale bids are calculated using the net wood cost less by-product sales.

Method 2: By-Products As Below-Line Income. Method 2 (illustrated in Figure 5.3) is a variation of by-products as a raw material cost offset. In this method chip revenue and other by-product sales are listed as an add-on, exclusive of sales and total operating expenses.

This method is frequently used by firms that are vertically integrated and use the by-products in-house. A significant quantity of timber from fee lands is usually used in the primary converting units (lumber and plywood plants).

Method 3: By-Products As Sales Return. Method 3 (illustrated in Figure 5.4) is a variation frequently used by companies that sell the primary product and the resulting by-products to outside sales firms. A significant portion of fee logs are usually used in the mill.

All outside sales are considered as a revenue rather than as a cost offset (Figure 5.2) or as a profit or loss (Figure 5.3).

The selected format is then used to document expected results by month and for the year.

Financial Exhibits and Trends

The completed pro forma profit (loss) statements are the basis for identifying trends and preparing other financial exhibits that provide a fuller understanding to the operator and to the firm's senior management. Five-year cost and performance trend charts or tables, performance and financial ratios and sensitivity analyses are some of the exhibits used. The actual format and content of the individual exhibits may change from year to year as the business or performance emphasis changes.

For example, a close tracking of defective material claims and the product mix may be important for the OSB producer during the early years of the plant. After the mill has developed a mature product and a stable, trained crew and as the equipment becomes older, the focus may shift to maintenance cost ratios.

The methods used in the forest products industry to achieve responsibility accounting have been summarized in this chapter. The operating budget—assumptions, cost detail and back-up information, and the resulting profit and loss summary—provides a yardstick . . . a yardstick to measure actual against expected profitability.

REFERENCES

Evans Product Co. 1982. *1981 Corporate Profile*. Portland, Oreg.
Hobbs, J. M., and Heany, D. F. 1977. Coupling Strategy to Operating Plans. *Harvard Business Review* 55(3):122.
Pyhrr, R. A. 1973. *Zero Based Budgeting*. New York: Wiley.

Six
The Computer As a Decision Aid for Management

There was a time when a sharp pencil, a well-oiled calculator and a swift office manager were about all that was needed to provide information for the decision maker. Those were the good old days when logs were cheap, labor plentiful and the sales price guaranteed a profit. Times have changed and profits are elusive.

The forest products manager requires timely and comprehensive information, accurate information that can be used to:

- Optimize the yield from the tree and the total resource
- Optimize mill profitability through decreased costs and improved product mixes
- Maximize man-hour and equipment efficiency
- Standardize and automate the process

Terms such as *optimize*, *maximize* and *standardize* imply the use of computers.

THE COMPUTER: FEATURES AND COMPONENTS

Computers come in all sizes and shapes. They range from the mainframe model designed to manipulate vast quantities of information to the microcomputer, which has at its heart a ¼-in.-square silicon chip etched to provide an intricate electrical circuit.

The computer can assimilate huge quantities of numerical information, digest and manipulate this information and present the facts or the action steps using the parameters designed by the user. All computers have four features in common: (1) an input function, (2) a storage function, (3) a program for processing the input information and (4) an output format. The output format will range from a computer printout to a full array of sophisticated machine commands. The latter will be discussed in Chapter 14.

Computers have two major components, the hardware and the software that is designed to drive the hardware. The hardware includes all of the mechanical and electronic components of the system; it is the physical, tangible items that the user can see and touch.

The second major ingredient, the software, is stored in the computer's memory and exists only as a programmed impulse. The structure of the program may range from simple to complex, depending on information needs. Creation of a complex program is complicated and labor intensive; the program may cost more than the hardware.

The actual hardware and software components are increasingly more sophisticated; they are evolving at an exponential rate. Each new system is more efficient and has greater utility and capability than earlier models. This increased capability permits the design of even more sophisticated computer systems.

COMPUTERS AND THE DECISION PROCESS

Each business or manufacturing process has both an external and an internal business environment. The cyclic, auction features of the forest products markets make the external business environment largely uncontrollable compared with a company's internal environment. The computer blends and harmonizes the two. The computer also harmonizes the planning, managing and decision-making tasks of the manager.

A computer has become a decision tool for management. (Courtesy *Forest Industries* magazine)

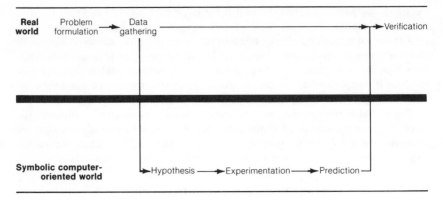

Figure 6.1. Decision making with models. (Source: Riggs 1968, p. 13)

"A manager's planning system serves as his interface with the external environment, while his control system provides and interfaces with the internal environment. It is unfortunate that some wood products managers use systems based on traditional procedures that do not fit today's world" (Bryan 1982, p. 1).

The manager's task is to manage resources: raw materials, human beings, production facilities, markets for products and available funds. The challenge is to optimize the return on these resources . . . variable resources that are part of an equally variable operating environment. Time often limits utility of the resources.

"The use of computers to aid in management decisions requires a very well defined purpose and continued focus on that purpose. A manager's purpose, whether he is in sawmilling or any other industry, is *to achieve the maximum return on his investments through the best assignment of resources under prevailing conditions.* It is frequently easy to lose sight of this purpose in the confusion and disruptions of day-to-day business activities" (Bryan and Lang 1974, p. 104).

Bryan (1982) describes the difference between current and should-be profits as the "profit gap". The gap represents the potential yet to be achieved; the gap can be closed in one of two ways. Figure 6.1 illustrates the options. The manager can make decisions and implement them in the real, everyday world to see what the results are, or he can test the decisions in a symbolic, computer-oriented world. The computer serves as an abstraction of the real world.

DATA: FUEL FOR DECISION MAKING

Data acts as the basis for decision making—the fuel that drives the computer program. Good-quality data, generated from both external and internal sources, are essential for smooth and accurate functioning of the program.

Internal sources of data may include historical information collected at an operation. Specific data are usually available by machine center or department. In addition, special studies are frequently required to identify specific facts that are not readily available. Miniprograms may be developed to organize the information into a useful form. Day-to-day control reporting plus data compiled as a result of a formal short-interval scheduling system make up a valuable data base.

The data acquired from a short interval scheduling system are particularly valuable. They not only detail current performance but also allow the analyst to forecast expected performance. Short interval scheduling is a structured performance measurement tool that identifies actual performance versus should-be performance based on defined time intervals, usually periods of two consecutive operating hours.

External data are obtained from a myriad of sources, which may include government, industry, research organizations sponsored by the industry, and industry observers. Computerized information retrieval services, such as the one available from the Forest Products Research Society (FRPS), are another important source. The FPRS retrieval service requires the user to provide key words; the service in turn provides a topical listing of literature by title and source. Either a one-paragraph abstract or the full article can be provided or located and the resulting information mailed or transmitted to the requester.

Trade associations are also important external sources of data. The American Forest Institute, Manufactured Housing Institute and American Plywood Assn. are some. The Western Wood Products Assn. is one more. The hardcover publication *Business Data and Market Information Source Book for the Forest Products Industry* (Dickerhoof and Kallio 1979) is an excellent guide to the external sources.

The external sources can be supplemented by analyst's reports periodically available from brokerage houses plus the published annual reports and 10-Ks of individual companies. Industry periodicals are another important source. In addition, consultants and resource persons can provide tips for locating more specific unpublished materials such as in-house reports.

IDENTIFYING THE QUESTION

The question is the trigger that activates a chain reaction of hardware, software and data utilization. The question has a twofold value. First, it clarifies the parameters for obtaining a satisfactory answer. Secondly, it provides the key for determining the computer technique that is best suited for the intended application.

Questions will focus on two areas. One type will deal with raw material selection and use; the other, with the manufacturing environment. The overriding consideration is how the external environment will affect the internal environment. The following types of questions lend themselves to computer analysis:

- What is the optimum allocation between manufacturing alternatives; that is, peel, saw, chip, flake or sell to an outside party?
- What is the conversion return for each log type or segment under the alternatives listed above or combinations of those alternatives?
- What is the break-even level for bidding on a specific timber sale?
- What falling and bucking practices will optimize the value of the log at the landing?
- What is the predicted merchantable volume by timber type using a sample scale?

- What is the optimum mix of log types and veneer purchases into a plywood plant?
- What breakdown pattern at the headrig will optimize the value from the log?
- What is the optimum mill design? How is this design conceived and what are the tradeoffs between equipment alternatives?
- How will a project be brought to a successful conclusion given an objective and limited resources?
- What is the best operating strategy given available resources and current market demand?
- What is the should-be operating schedule and work center assignment by day, shift or other production cycle?
- What are the optimum inventory practices for logs, supplies, in-process goods and finished products?

These are a few of the possible questions. Each operator will develop his own based on the specifics of the operation.

MANAGING THE SOFTWARE TOOL

In the early days of computer little attention was paid to the format (or form) of the answers and supporting information that the computer produced. The manager had to rely on the computer technician to translate the mass of data.

Commented one plywood manager after a confusing experience with a computer printout and the technician, "He didn't understand plywood and I didn't understand how the answer was arrived at and he couldn't explain it to me."

Frequently a computer printout will provide so much data that the answer is obscured in a mass of figures. Add numeric symbols and codes, and the resulting information can overwhelm the user. The manager and the technician should decide the format for the answer beforehand, to ensure that the summary sheets and related tables and data produced will provide a clear trail that identifies and supports the rationale, clarifies the specific facts and identifies the operating guidelines for achieving the answer in the real world. The following are further guidelines for managing the software tool.

Select a coordinator who understands both the business and computers. Then select a team to represent the various business facets, such as operations, industrial engineering, finance and accounting, and others as appropriate. This team will provide the input as the model (the program) is constructed and act as an ongoing resource to the coordinator. Their participation will build a commitment to testing and implementing the program results.

The results may trigger a "What if . . . ? response. A sensitivity analysis provides the opportunity to determine the effect on the output with varying levels of input. This analysis should be completed before the computer recommendations are implemented. The added risk and worth of incremental benefits may not be sufficient to justify the optimum course of action.

The model can be checked by using historical information. If the objective of the model is to predict sales, data from previous years can be used to predict last year's sales results. Agreement with real-world results will indicate the worth of the model and its assumptions.

COMPUTER PROGRAMS: PLANNING TOOLS FOR MANAGEMENT

Computer programs come in all sizes and degrees of complexity. Some are easily constructed with little more than a few simple subroutines. Others are complex, with seemingly endless reiterate routines. Some are designed to speed accounting chores; others, such as the network models used in detailed planning, are complex. No attempt will be made to describe these models cookbook style. The intent is to acquaint the reader with each as a potential management tool. One such tool is the linear program.

Linear Program

The linear program is increasingly being used as individual mills and integrated complexes upgrade from one-log, one-product operations. This model is sometimes defined as "an iterative mathematical technique for finding the best use of resources subject to constraints" (Dunn and Ramsing 1981, p. 25).

Linear programs are used to optimize resource use and improve profits. Log mixes, product mixes and other resource optimization problems are solved using this technique. The first step in setting up a program is to prepare an objective statement that identifies the goal of the program.

"In our business, where profit margins are getting smaller and smaller, we needed to know how to best utilize the logs in order to generate the highest possible contribution to profits. We needed to be able to select a feasible product mix and manufacturing strategy that would maximize contribution to profit, under specified conditions. And we needed to receive all of this information quickly, regularly and accurately" (De Guzman 1982, p. 31).

This manufacturer identified the need, originated the objective statement and identified the conditions, including a time constraint. The resulting profit-optimization model aided the management of this Philippines plywood plant in determining both log use by type and the best product mix, and it also assisted management in scheduling. The printout from this linear program identified areas of opportunity and validated courses of action prior to implementation in the mill. Figure 6.2 is a summary of the internal and external considerations in setting up a program for an individual plywood plant.

Figure 6.3 is an example of the typical profit and loss summary that results from each run. The volume by product category is identified along with the expected production levels, costs and sales returns. Backup data will provide the details on the input, mode of operation, output, costs and crewing levels. Also included in the backup data is the detailed production schedule by machine center; this details the machine hours necessary to achieve the planned results within the expected performance levels of employees and equipment.

Simulation Model

A simulation model is used almost as frequently as a linear program. One writer describes the process as follows: "A quantitative technique used for evaluating alternative courses of action based upon facts and assumptions with a computerized

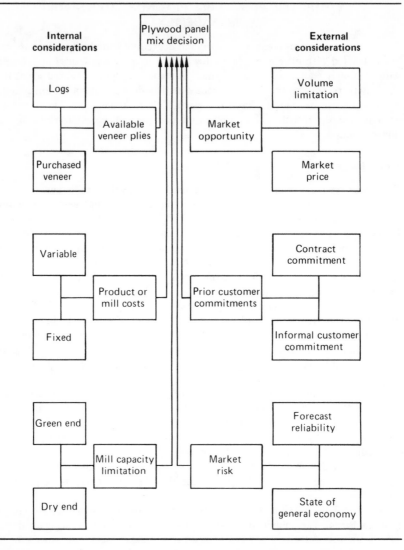

Figure 6.2 Internal and external considerations in the plywood panel mix decision. (Source: Baldwin 1981, p. 273)

mathematical model in order to represent actual decision making under conditions of uncertainty" (Thierauf and Grosse 1970, p. 471).

Simulation is reproduction of the operating environment. A simulation model mirrors the cause-and-effect relationships between raw material, people and machines. Often simulation models rely on some type of "generator," such as a random-number table, to develop the randomness that occurs in real life. The result-

Profit and Loss Summary
Oregon Atlantic Plywood

	M3/8	$ amt.	$/unit
Sales			
Sanded plywood	4,128	691,307.06	167.47
Sheathing plywood	8,566	1,098,311.00	128.22
Plywood siding	3,934	809,636.38	205.80
Misc. sales + adj.	16,629	−1,662.85	−0.10
Total sales	16,629	2,597,591.59	156.21
Total materials		1,499,902.00	90.20
Direct labor and P/R costs		258,126.44	15.52
Mfg. expense – fixed		128,925.00	7.75
Mfg. expense – variable		364,296.25	21.91
Cost of goods sold		2,251,249.69	135.38
Gross profit		**346,340.00**	**20.83**
Shipping expense		32,924.47	1.98
Selling expense		5,986.27	0.36
G + A expense		27,100.00	1.63
Total operating expense		66,010.74	3.97
Operating profit or loss		**280,329.31**	**16.86**
By-product income		*189,313.00*	*11.38*
Profit/loss before taxes		**469,462.31**	**28.24**
Recovery (M3/8 plywood/mbf)		3.15	

Figure 6.3 Sample profit-and-loss summary.

ing pattern of occurrences shows what the system will do within a given set of conditions. Unlike the linear program, the simulation model simply portrays the real world given a set of alternatives.

The technique is frequently used to test mill flow designs to identify the results of a given production schedule or raw material. It is ideal for assessing the detailed impact of new products, methods and machines. In addition, it is frequently used to model sales and physical distribution systems.

Waiting Line (Queuing) Model

Waiting line (or queuing) models, more specialized simulation methods, are used frequently as cost minimization techniques. These models are particularly suited for material flow problems, such as in-process lumber flow in a sawmill or truck arrivals at the log yard.

The techniques, which were originated by A. K. Erlang about 1909, are used to minimize the sum of congestion and service costs. The manager can then test out alternatives to arrive at the least cost option. A waiting line model is constructed in the following steps:

Step 1: Identify the problem. For example, the high-capacity and costly headrig is underutilized.

Step 2: Determine the objective. In this example the objective could be either (1) to find the most cost-effective operating policy given a desired level of output or (2) to identify possible modifications to the existing mill design in order to improve overall productivity.

Step 3: Gather the facts related to the system being studied.

Step 4: Test the possibilities using probability methods such as Poisson distribution or Monte Carlo random-number technique. Poisson distribution is a discrete probability distribution that predicts the number of arrivals at a given time. The Monte Carlo technique uses random numbers to create statistical distribution times.

Step 5: Implement and test the results.

The queuing technique continues to become more sophisticated. Rather than trial-and-error manipulation of each computer run, the operator can program the model to balance the expected costs of idle time and then compare those costs with the expected costs of waiting time. Achieving a balance of these two factors will provide the least total cost.

Network Models: PERT and CPM

"Who does what by when at what cost" is a question frequently asked by effective forest products managers prior to committing resources. A network model, designed to schedule resources to meet an objective, can be an excellent tool.

"I knew that raw materials, engineering and marketing studies plus extensive financial analysis indicated that we should convert our board plant into a waferboard plant. We had the resources. We just had to figure out how to fit the project together," commented a project manager.

As have others, this manager has found that computerized resource scheduling programs such as PERT (Program Evaluation and Review Technique) and CPM (Critical Path Method) are invaluable management tools. Each of these techniques is available in canned computer programs or a program can be constructed to suit the specific situation. The scope of use can be limited to the construction portion of the project, or it may include log procurement, marketing activities and all other activities as part of overall project management.

These network models, best described as graphical representations of project activities, are an extension of the Gantt system, but with a more sophisticated analysis of activities, times and resources. PERT and CPM, developed about 1958, have a lot in common. Each does the following:

- Breaks the project down into component activities
- Places a restriction on each activity by identifying what immediately precedes it, what immediately follows it and what other activities can be done during the same time
- Constructs a network that is a graphical expression of the activity list and the restriction list

The main difference between PERT and CPM is that the latter uses a single time estimate for each activity while PERT has three probabilistic time estimates: optimistic, most likely and pessimistic. Each network model identifies the critical path, the longest path through the network for which there is no slack time. A rough rule of experience indicates that only about 10% to 15% of the activities for a project will be critical at any one time.

Both network models have the following advantages for the forest products manager. They

- Aid management planning by reducing the number of activities that may be "forgotten" in planning and developing large-scale projects.
- Provide a check on the basic logic of the plan by requiring a systematic "think it through" process.
- Are easy to use and understand. They provide coordinated control between job functions, with each participant knowing what is expected by when, by whom and at what cost.

Line of Balance Analysis

An additional management tool is the line of balance analysis. This technique is used to determine actual progress versus the schedule. It determines the point at which all activities should be on a given date and compares this with the actual status of each activity on that date.

Network, waiting line, simulation and linear programming are only a few of the many techniques for planning. Figure 6.4 summarizes the question type for each of these models. Each year more sophisticated models are finding wider use in the forest products industry.

Figure 6.4. Computer models: the question each type answers.

Model	Question type
Linear program	What is the most profitable product mix from the log?
Simulation	What effect will a manufacturing strategy have on the mill or business?
Waiting line/queuing	How can I improve lumber flow and increase production in the sawmill? Where are my bottlenecks?
Network (PERT/CPM)	How can I best schedule resources to meet a project objective?

While planning models are used extensively by management, no less important are control programs that are individually tailored for each mill. These include:

Sales and inventory tracking
Mill production scheduling
Product mix summaries
Cost analysis and forecasting
Historical recap of customer purchases

The value of the computer cannot be underestimated. The 1960s witnessed its use as an automated bookkeeper; the 1970s saw its increased use in planning, with some machine control functions gradually being adopted. The remainder of the century will witness the marriage of tracking, planning and machine control functions into an interrelated computerized business and manufacturing system. The forest products operator who readily accepts this innovation will have the competitive edge in obtaining raw material and markets and will close the profit gap.

REFERENCES

Baldwin, R. F. 1981. *Plywood Manufacturing Practices*. Rev. 2d ed. San Francisco: Miller Freeman Publications.

Bowyer, J. L., and Carino, H. F. 1982. Sawmill Analysis Using Queuing Theory Combined with a Direct Search Optimizing Algorithm. *Forest Products Journal* 31(6):31.

Bryan, E. L. 1982. Microcomputer Applications for Wood Products Management. Reprint of an address presented at the North American Sawmill and Panel Clinic and Machinery Show, March 3–5, 1982, Portland, Oreg. San Francisco: Miller Freeman Publications.

Bryan, E. L., and Lang, J. R. 1974. Computerized Planning for Maximum Profit. *Modern Sawmill Techniques*. Vol. 3, p. 103. Proceedings of the Third Sawmill Clinic, Portland, Oreg., February 1974. San Francisco: Miller Freeman Publications.

Bryan, E. L., and Ridgeway, R. E. 1979. Factors Affecting Plywood Profitability: Logs, Equipment, and Product Mix. *Modern Plywood Techniques*. Vol. 7, p. 43. Proceedings of the Seventh Plywood Clinic, Portland, Oreg., March 1979. San Francisco: Miller Freeman Publications.

De Guzman, J. M. 1982. Operation Improvements Are a Key to Profitability. *World Wood* 22(2):30 (February).

Dickerhoof, E., and Kallio, E. 1979. *Business Data and Market Information Source Book for the Forest Products Industry*. Madison, Wis.: Forest Products Research Society.

Dunn, R. A., and Ramsing, K. D. 1981. *Management Science: A Practical Approach to Decision Making*. New York: Macmillan.

Riggs, J. L. 1968. *Economic Decision Models for Engineers and Managers*. New York: McGraw-Hill.

Thierauf, R. J., and Grosse, R. A. 1970. *Decision Making Through Operations Research*. New York: Wiley.

Seven
Resource Management and the Changing Forest

Years have passed and so have the giant trees. The stately eastern white pines, the giant Douglas fir, the mammoth redwoods and the beautiful fine-grained southern pines are about gone. The forest resource is changing.

Product performance and appearance characteristics have become more important as the size and concentration of the big-tree timber stands has dwindled. Species such as the hemlock and the true firs, once considered weed trees, are now filling the resource gaps.

Ponderosa pine in the Black Hills in South Dakota, live and dead lodgepole pine in southwestern Wyoming, western larch in the interior of British Columbia, balsam fir in New England and pond pine in the Carolinas are additional species that are taking up the slack.

Other suitable softwood growing sites around the world are producing logs for the industry—the red pine from Scandinavia and the radiata pine of Australia and New Zealand. These and other small-diameter species now dominate the softwood forest of North America, Europe and across Russia onto the islands of Japan.

Logs with natural defects such as shake, spiral grain and sweep are increasingly recognized as opportunities rather than problems, as scientific and electromechanical state-of-the-art technology deals with the diversity of size, defect and species.

Professional resource management has followed in step with state-of-the-art technology. Some resource management groups have even introduced a species, managed it to maturity and then introduced the resulting tree to the wood-using industry. Large-tree growing areas have become small-tree growing areas as the fiber needs of a nation shorten the growing cycle. The demand for wood fiber increases as the harvest age decreases. The situation in the USA is representative.

In 1976, the total U.S. demand for wood was 13.3 billion ft^3; this represents a sharp increase from a decade and a half earlier. During that same period, the average log diameter decreased sharply; the species and log types utilized increased sharply. Small-diameter species now dominate the log yards of the converting mills as the industry spreads into the northern regions and the higher altitudes.

Small-diameter species and small logs mean different things to different people. In eastern Canada, Scandinavia and the southern USA logs are considered small when the diameter ranges from 4 to 10 in. In western Canada, the Pacific Northwest and the Inland Empire region of the USA, as well as in the plantation pine areas of New Zealand and Australia, the diameter of a small log may reach upward to 20 in. In some areas of the world a log is considered small if it is under 30 in. The actual definition is relative, relative to the size of the total resource or in comparison with the logs of an earlier era.

Mammoth Douglas fir peelers are about gone. (Courtesy Simpson Timber Co.)

Small-diameter species are now filling the resource gap.

SMALL LOGS: PLUSES AND MINUSES

There are certain inherent advantages in utilizing small logs. They enhance an automated process; they are uniform in size and have other homogeneous physical characteristics. The prime advantage is availability; the disadvantage is that the mill becomes increasingly volume sensitive as the diameter shrinks. Sixty logs per shift has increased to 2,000 and more in some prime producing areas. The design and operation of a small-log mill relies more on skilled engineers, mathematicians, programmers and electronic technicians than on the skilled sawyers or lathe operators of the past.

"It used to take five years or more to make a sawyer on a headrig," cited a senior manager. "We qualify an operator in thirty days or less with all this fancy hardware. All the operator has to do is watch the panels and push the buttons . . . occasionally."

Less reliance is placed throughout the operation on manual effort and more on mental effort. The primary converting plants, such as lumber and plywood operations, are going through a major evolution. The rate of change has been slow but is accelerating at an exponential pace.

BASICS FOR SURVIVAL WITH THE CHANGING RESOURCE

Demand for forest products has drawn heavily on the available log supply. Logs have soared in price and become 60% and more of the finished-product value.

Uniform logs enhance the automated process.

Therefore, effective utilization and accurate measurement of the log—and a carefully negotiated price based on these two considerations—are essential for survival.

Improving Fiber Utilization

Only about 75% of the developing fiber actually leaves the forest. Logs, limbs, tops and residual log segments are left to be burned or to decay. The logs are taken to the mill, where 40% to 50% of the fiber is lost to residuals such as sawdust, slabs, shavings and other by-products.

One writer cited the importance of only a 1% improved utilization for one state: "Basically it is assumed that better utilization means more wood is delivered to Arkansas mills or that fewer acres need to be harvested to satisfy mill demands. . . . A one percent improvement in utilization could translate into 376 jobs and 2.5 million dollars to the . . . economy" (Porterfield 1976, p. 8). These benefits are in addition to the additional forest products that would be available for the consumer.

There are several factors inhibiting improved utilization. One is harvesting methods and the cost of using these methods; slow acceptance of treelength or whole-tree logging is another. Systems of scaling and measurement are yet another. The last factor is important; it is the overriding consideration for the logging contractor and the mill. Economics weigh heavily in the decisions of each.

Standardizing Fiber Measurement

Commented a private industry source: "It is fundamentally impossible to grow small-diameter timber, and handle it at a profit, if we have no accurate method of measuring what we are doing."

The situation facing the buyer and seller may be compared to trying to time a runner over a set distance, with someone forever moving the start and finish posts. The solution is an improved system of measurement; the start and finish posts must be fixed. The starting post for the forest industry manager is an accurate and consistent scaling system. A meaningful unit of measurement must be consistently applied to logs of all lengths and sizes.

History of Log Rules. Log rules or systems of measurement come in all types and variations. The Mercer Tables are used in Pakistan; the Canterbury Tables are used in New Zealand. The Scribner C rule is used in the Pacific Northwest, with the Doyle rule used extensively in the southeastern USA. Ninety and more board-foot log rules have been developed and put into use throughout North America alone since the first rule was published in 1825. In addition there are countless local variations that are intended to overcome a perceived shortcoming in an existing rule. Fortunately all have not survived.

The surviving rules and variations are the result of differing utilization standards and associated allowances for slabs, taper, shrinkage and saw kerf. The calculation method or formula is another variable. There are three methods by which log rules are prepared:

Mill tally: The lumber developing from straight defect-free logs or logs typical of the resource base is recorded and tallied for each diameter increment. Increments of length may be an added variable.

Diagram rule: The best known example of this method is the Scribner rule and its many variations. Each small-end diameter increment is diagrammed into 1-in. boards to identify the potential lumber content of the log. Saw kerf, shrinkage, taper assumptions, sawing patterns and other variables are factored into the diagram.

Log scaling and volume determination provide an important scorecard.

Formula rule: A formula is developed to represent the log as a geometric shape. Adjustments and allowances are made for saw kerf, edgings, slabs and other milling losses.

A rule may be prepared using a combination of methods. Each rule answers a need within a set of prescribed conditions.

As the industry shifts to small logs and diverse species, the prescribed conditions may no longer fit a significant segment of the log population. It becomes increasingly difficult to place a value on a log when the recovery of primary product decreases rapidly below 10 in. Commented an industry spokesman, "A mill can go bankrupt with a thirty percent overrun if it is cutting only eight-inch logs, but would make a profit at thirty percent while cutting eighteen-inch logs."

Cubic Measure. Some industry veterans advocate a universal log rule; others, such as the large landowners, recognize that small logs will increase in quantity and that an accurate cubic formula is needed to determine the actual fiber developing from their lands. Others see the cubic formula as a better way to buy and sell small logs, since small logs are frequently used for purposes other than lumber manufacture.

The British Columbia government, with extensive provincial lands, has adopted the cubic measure. Weyerhaeuser and Champion Intl. are among other major landowners that use the cubic formula. The cubic measurement has several advantages. It

1. *Measures all the fiber in a log* rather than just the assumed lumber yield. Cubic measurement is a rational accounting tool.
2. *Allows easy comparison of log allocation alternatives* between competing conversion options. Cubic measurement lends itself to sophisticated management decision-making techniques that seek the most profitable combination of products.

Figure 7.1. The Smalian formula. (Source: Baldwin 1981, p. 60)

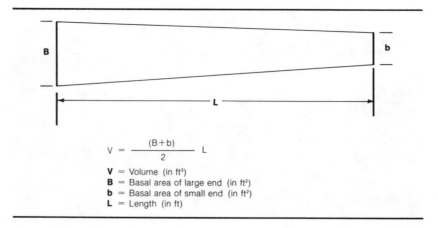

$$V = \frac{(B+b)}{2} L$$

V = Volume (in ft³)
B = Basal area of large end (in ft²)
b = Basal area of small end (in ft²)
L = Length (in ft)

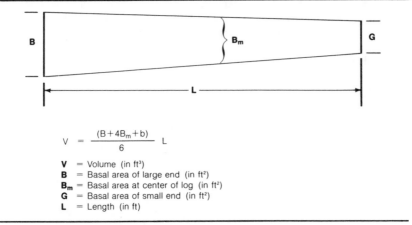

$$V = \frac{(B + 4B_m + b)}{6} \, L$$

V = Volume (in ft³)
B = Basal area of large end (in ft²)
B_m = Basal area at center of log (in ft²)
G = Basal area of small end (in ft²)
L = Length (in ft)

Figure 7.2. The Newton formula.

3. *Simplifies yield comparisons, encourages technology transfer and makes comparative data available* on a regional and national basis. Cubic measurement can be used as a worldwide communication vehicle when combined with simple metric conversion factors.

The cunit, 100 ft³ of solid wood, is the accepted unit of measurement. Cubic rules are based on mathematical formulas; the most widely used formula is the Smalian rule. This rule is a compromise between accuracy and ease of measurement. Figure 7.1 illustrates this rule. Simply stated, the area of the large end of a log segment and the area of the small end are summed, the sum is divided by 2, and the number is then multiplied by the length of the log to produce the cubic measurement.

There are other rules, such as the Huber, the Sorensen and the Newton. The last is generally accepted as being the most accurate but is also considered the most difficult to use in an industrial setting. Figure 7.2 shows the formula.

SAMPLE SCALING:
THE APPROACH FOR SMALL LOGS

The cubic rule and small logs have combined to further another industry trend, the sample scale. As the piece count has soared with the trend to small logs, scaling of individual logs has become increasingly impractical. Sample scaling has been determined to be a surprisingly accurate method to measure a homogeneous small-log population using piece count and weight as the two key variables.

Generally, the total number of loads to be removed from the site, the desired accuracy, and the coefficient of variation are the ingredients in determining the sampling frequency. Total volume, species mix, sales value and log count all play a part in determining the coefficient of variation.

An accurate sample scaling system requires a statistically valid sampling plan for each distinct timber-growing area. The data generated, by log and species type, are then projected to reflect the entire source population with the aid of a regression formula as part of a computer program.

Sample scaling has a number of advantages; it

- Reduces the operating costs for scaling at the landing or log yard without reducing control
- Enables the operator to handle the increased piece count with small logs more efficiently
- Increases report availability while reducing data processing costs

Combined with cubic measurement, sample scaling is an important part of a utilization control system for the manager.

An accurate scaling system aids the manager to recognize utilization possibilities; but profit-increasing opportunities hinge on the following:

- Recognition that the traditional large log is about gone. Small-log species and defective material will make up a larger portion of the furnish.
- Selection of merchantable trees from the cutting site must include all portions of the wood basket that can be profitably harvested. This includes tops, segments and other portions of the tree that have been heretofore left in the woods
- Realization that the traditional board-foot log rule, based on expected lumber outturn, is becoming less useful as conversion options increase and the character of the log changes.

And last of all, the manager must utilize the log in an increasingly higher value combination of products. Each incremental increase in utilization then becomes the base for establishing additional attainable goals.

REFERENCES

Baldwin, R. F. 1981. *Plywood Manufacturing Practices*. Rev. 2d ed. San Francisco: Miller Freeman Publications.

Clark, A., III. 1978. Total Tree and Its Utilization in the Southern United States. Paper presented at Session 16, 31st Annual Meeting of the Forest Products Research Society. *Forest Products Journal* 28:10.

Dilworth, J. R. 1977. *Log Scaling and Timber Cruising*. Corvallis, Oreg.: OSU Book Stores.

Fahey, T. D., and Starostovic, E. 1979. *Changing Resource Quality: Impact on the Forest Products and Construction Industry*. Madison, Wis.: Forest Products Research Society.

Fahey, T. D., and Woodfin, R. O., Jr. 1976. The Cubics Are Coming. *Journal of Forestry*, November, p. 739.

Freese, F. 1973. *A Collection of Log Rules*. USDA Forest Service General Technical Report FPL 1. Madison, Wis.: Forest Products Laboratory.

Porterfield, R. L. 1976. Improved Utilization—Where Do We Stand? Paper presented at the annual meeting of the Mid-South Section of the Forest Products Research Society, Jackson, Miss., October 13–14, 1976.

Texas Forest Service. 1979. Log Rules Versus Lumber Value. *Forest Products Notes*, vol. 4, no. 10. Lufkin, Tex.: Forest Products Laboratory.

Williston, Ed M. 1981. *Small Log Sawmills: Profitable Product Selection, Process Design and Operation.* San Francisco: Miller Freeman Publications.

Eight
Value Management:
A Log Allocation Concept

There was a time when logs were felled, bucked and yarded with little regard for potential value. Felling and bucking instructions were simple; prime segments were cut into peeler multiples for the veneer plant, with the remaining portion either cut in 2-ft increments for the headrig or left to rot in the woods.

Utilization standards were low; it was not uncommon for a third or more of the log volume to be left behind in the woods. The typical Pacific Northwest crews, frequently bushlers or gyppos, were paid a piece rate for the log scale tallied. A crew had little time for small logs, other species or difficult-to-yard segments or chunks.

The U.S. Southeast was little different. Product selection was merely a matter of selecting the larger, defect-free logs for the sawmill. The remaining logs, small-diameter or larger logs of poorer quality, were left in the woods to be harvested by a shortwood crew.

Mill utilization was little better. The sawyer's job was to fill the cutting orders; anything more than that was secondary. Hard times, expensive stumpage and dynamic industry leadership are changing all that.

One industry leader set the pace. George Weyerhaeuser, during a speech in the mid-sixties, described the log as a product—a primary product that is the prime ingredient for serving myriad consumer needs. He defined the corporate goal of Weyerhaeuser Co.: to maximize the net return by allocating the log to its optimum combination of uses.

Land and timber stewardship was called high-yield forestry; the log allocation yardstick is called return to log (RTL). The value management concept, an idea achieving ever greater acceptance, is a radical departure from the traditional single-product concept—the practice of matching the log to the mill to produce a product a customer will buy at some price. Variations of the value management concept are called conversion return, log value analysis as well as a variety of other terms.

VALUE MANAGEMENT—WHAT IT MEANS

Under the value management concept, a specific log or log type will be manufactured into a variety of products. The products chosen will be those that yield the most profitable return. A systems approach is used, from the land base through to the customer.

The forest industry manager, under the concept, must no longer think as a single-product converter but will manufacture any product that will provide the highest net return from the log.

The concept requires organizational flexibility. The traditional compartmentalized forest products organization is often too rigidly structured and formalized to have the expertise and the flexibility required to meet both short-term and long-term cost and value objectives.

ASPECTS OF VALUE MANAGEMENT

Land Management Practices

The character and volume of the resource will be identified; the species, the timber profile, the defect characteristics, the character of the wood (slow growth versus fast growth, for example) and the volume per acre will be documented. Harvest restrictions, such as site and weather constraints, will be identified.

The manager will recognize that productive forest lands are becoming a relatively scarce resource and must be used efficiently to meet the ever-increasing demand for forest products. A scenario of reforestation, growth, harvest scheduling and overall management will be developed.

The lands will be managed for both short-term and long-term objectives, recognizing changing customer demands and preferences. In addition, the manager will make planning assumptions regarding future application of forest management practices and timber utilization standards.

The Conversion Process

The log or treelength stem will be processed into products that fill a variety of needs in the marketplace. The manufacturing process will cease, and the product will be sold when incremental processing would result in a diminishing return to the balance sheet. Solid-wood products such as lumber and plywood may be produced; other portions of the log may yield the raw material for reconstituted products such as flakeboard or medium-density fiberboard.

Leftovers may be used in digested fiber products such as paper or as a replacement for a fossil fuel. Port Orford cedar may be exported as whole logs rather than sawn; rough green 1x4s may be sold to a specialized converter rather than kiln dried and surfaced by the manufacturer.

The operator must merchandise the log and process the resulting segments with well-controlled manufacturing practices. The mills will be designed for the tree, rather than the traditional approach of selecting the log for the mill. The operator must have an intimate knowledge of equipment types and the various yield and product-value trade-offs that will occur within a specific log type. A knowledge of

Under the value management concept, lands are managed for both short-term and long-term objectives.

computers and sophisticated control equipment is becoming increasingly important to the operator.

The greatest incremental value and the most efficient use will be achieved as the mill manager seeks a higher recovery of primary product. Overall, management will recognize that efficient and cost-effective utilization not only makes the process more profitable but adds immediate value to the fee stumpage or other timber acquired.

Log Allocation

The manager will have a detailed knowledge of the product options available from a log type. In addition, market demand and expected prices will be known or forecast. For example, single-item pricing is usually more profitable than Standard & Btr pricing of dimension lumber. The former requires more customer knowledge and better marketing skills, but the overall sales return is usually higher.

Log segments will then be allocated to the various conversion options based on current or expected market demand. When a number of log types, species, conversion facilities and product options are available, the actual allocation decision is usually based on a sophisticated log allocation model.

The allocation decision must then be reevaluated whenever a significant change occurs in the timber base, the conversion process, the cost base or the market demand. Short term, the existing conversion facilities may constrain the allocation model; long term, the strategic planning activity (Chapter 4) will include an allocation model as a key ingredient in determining future facilities.

LOG TESTS: THE FOUNDATION FOR THE ALLOCATION DECISION

Yield analysis, or log tests, provides the facts for the allocation decision-making process. The log population for a test may be one or more of three specific types.

Identifying the Test Population

Camp Run Batch. A mix of log grades and diameter classes from a specific sale, vendor or geographic area is selected for a camp run batch test. The test results are useful only if the test population represents a typical log that is easily defined by both the supplier and the mill. A camp run batch test is frequently used to establish a sample scaling system (discussed in Chapter 7). Mill yield studies are used to validate the accuracy of the sample scale over time.

Log Grade Sample Within Specific Diameter Classes. Grade sampling within diameter classes is the most frequently used test type for log allocation decisions. The log sample should include diameter increments that correlate with a log rule and provide a noticeable yield change.

Log tests provide key information for the log allocation process.

For example, the No. 3 peeler log criterion specifies a 24-in. minimum diameter; No. 1 and No. 2 peelers have a 30-in. minimum diameter. A No. 3 peeler under 30 in. may have No. 1 peeler characteristics, with the resulting veneer grade yield higher than that from a No. 3 peeler that is 30 in. or more in diameter. A 6-in. increment is usually satisfactory for a log grade subcategory for the old-growth regions.

Test samples may be in 1-in. increments as the diameter drops below 20 in. For example, the product yield of lumber and veneer can decline 7% or more for each 1-in. decrease in log diameter below 10 in.

Individual Logs. Data obtained from testing individual logs provides the most detailed and comprehensive information. Individual logs are selected as a representative sample of a particular log type, diameter class, species or log characteristic. Although previously used occasionally for U.S. Forest Service and research-oriented tests, this method of testing individual logs is being used more frequently to validate the return on investment for the new generation of sophisticated electronic process control equipment.

Selecting the Log Type

When selecting the log type for initial tests, that type representing the greatest volume of logs should be given prime consideration. As a rule of thumb, the larger the population, the greater the need for ongoing testing prior to the periodic allocation decisions.

Logs representing a small yet significant portion of the actual or potential furnish should be tested as individual logs within a larger batch test of like material. The range of results and the norm to be expected will then be known.

The volume of the sample should be a compromise between as large a volume as possible and the capability to track the test volume through the mill. A 2 to 4-hour test run is usually sufficient to represent the log type being tested.

Conducting the Test

After the study sample is selected, the logs should be check-scaled to ensure accuracy. Those operations using a sample scale should correlate the check-scale volume with their sample scale results, usually using weight and piece count criteria, to determine the accuracy of these sample scaling procedures.

The logs are then tagged or otherwise identified and isolated from the ongoing log inventory to facilitate tracking during subsequent handling, barking and bucking into segments. A comprehensive description of individual log characteristics will aid interpretation of the test results.

The test objective is achieved when all the primary product, such as lumber or plywood, and the residuals produced from the log test are accounted for.

The key measurement and data collection points in the manufacturing process are schematically represented in Figures 8.1 and 8.2: the former, a typical production flow for plywood; the latter, for lumber.

Plywood. In Figure 8.1 the flow begins as the log is received; it ends as the by-products exit and are tallied and as the primary product is prepared for shipment.

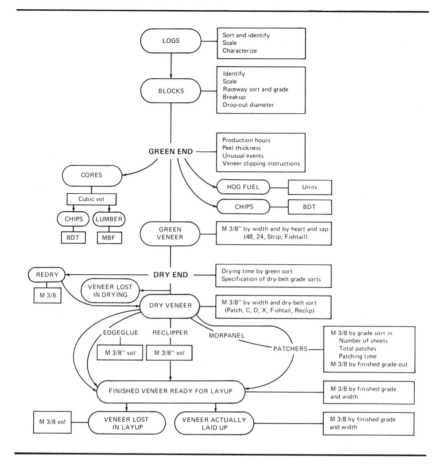

Figure 8.1. Plywood log test process flow. (Source: Baldwin 1981, p. 53)

The flow may vary depending on the log type consistently used in the region and the panel type being constructed. A sheathing mill in the South will have few steps compared with a West Coast grade mill utilizing a cold-peeled cull log.

Lumber. The flow in Figure 8.2 begins as the log is identified in the sort yard or pond. Primary product and by-products develop at each primary and secondary breakdown unit in the sawmill. The process flow illustrated is general; as with plywood, the actual flow will depend on log type, equipment configuration and volume expectations through the mill.

In a log test minor volumes of by-products may be accounted for by previously developed and validated waste factors. This procedure is a useful tool when the expected waste factors remain relatively constant among tests. A log sample disproportionately weighted to either end of the spectrum—from defective material to sound logs—will make this shortcut impractical.

Recording the Results

The information from the log test should be presented concisely and in an easily understood manner. Figures 8.3 and 8.4 are example one-page summaries for each primary product tested.

Figure 8.2. Lumber log test process flow.

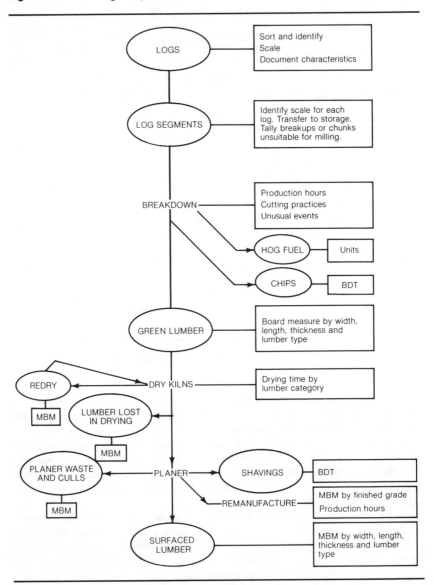

Log Type: No. 3 Douglas Fir Sawlog, Highline Classification

	Log diameter (in.)			
	8	**9**	**10**	**11**
Rough green lumber recovery (MBM/ccf)	.792	.850	.891	.877
Realizable value ($/ccf)[1]	175	196	206	202
Grade yield (% of total)				
Select	28.2	31.5	29.2	23.1
Construction (#1)	43.2	39.8	42.9	44.1
Standard (#2)	13.4	16.8	14.8	16.3
Utility (#3)	14.7	11.2	12.3	14.1
Economy (#4)	.5	.3	.8	2.4
	100.0	100.0	100.0	100.0
Length distribution (% of total)				
16′	66.6	75.6	78.9	67.8
14′	13.2	8.1	6.2	17.2
12′	12.8	11.3	8.3	8.2
10′	2.1	1.9	1.1	2.7
8′ & shorter	5.3	3.1	5.5	4.1
	100.0	100.0	100.0	100.0

[1]Primary product value (manufacturing costs and by-product revenues not included)

Figure 8.3. Example log value analysis: lumber.

Figure 8.4. Example log value analysis: plywood.

Log Type: No. 3 Douglas Fir Sawlog, Highline Classification

	Log diameter (in.)			
	8	**9**	**10**	**11**
Net veneer recovery (M3/8/ccf)	1.32	1.43	1.46	1.62
Veneer in-panel value ($/ccf)[1]	212	230	238	265
Grade yield (% of total)				
AB	5.0	6.0	6.5	11.0
C	87.1	85.2	78.1	69.2
D	7.9	8.8	15.4	19.8
	100.0	100.0	100.0	100.0
Width distribution (% of total)				
54	55.0	57.0	65.0	73.2
27	10.3	11.2	10.6	8.3
Random widths	30.9	28.3	21.7	16.3
Fishtails	3.8	3.5	2.7	2.2
	100.0	100.0	100.0	100.0

[1]Primary product value (manufacturing costs and by-product revenues not included)

The two figures represent comparative tests using a like log: 16-ft lengths for the lumber test run and 17-ft logs for a comparative plywood run. Four 1-in. diameter increments, 11 in. and below, were tested using a so-called Highline No. 3 Douglas fir sawlog typical of the second-growth stands on the eastern fringe of the Willamette Valley in Oregon.

Figure 8.3 summarizes the rough green lumber recovery, the lumber value per cunit, the grade yield and the length distribution. This information plus other yield and cost information provides the basis for performance comparisons within mills, between equipment types (such as a Chip-N-Saw versus a Maxi-Mill) or between competing uses for the same log, such as lumber or plywood manufacture.

Figure 8.4 is the plywood counterpart using the same log type as in Figure 8.3. This summarizes the net recovery, veneer in-panel value, grade yield and width distribution for the four diameter classes.

Analyzing the Results

Figures 8.3 and 8.4 illustrate the comparative statistics for this log type. Each figure is a summary sheet that identifies the product recovery, the grade yield and the character of the product. Although manufacturing costs and by-product revenues are not included, each is an important part of the analysis.

Short range, the results can be used to allocate the log type between existing competing conversion units. Long term, the information can be used to identify products, equipment, and manufacturing strategies that may be required to maximize the return to the stump or that make it more profitable to manufacture an alternate product.

This analysis method is not reserved for plywood and lumber testing only; testing may indicate that the net return is higher for reconstituted boards such as oriented strand boards rather than veneer (plywood) or lumber manufacture. It may also indicate that a different conversion option or product type is needed to justify conversion into a forest product. This was particularly true in Canada and later in the Upper Midwest when waferboard plants were sited to justify the conversion of a heretofore weed species, aspen, into a solid wood product.

Note that Figures 8.3 and 8.4 use the cubic measure as the common denominator. This common denominator provides a useful tool for evaluating the return per unit of logs for each conversion option.

Log testing is neither easy or free; a test may be simple or complex. The test objective and the design of the test plus the training of the participants will determine the cost or complexity. Testing generates base data that can be used by the operator to select the manufacturing option that will yield the greatest net value; the data can also be used to establish yield cost and performance standards for a specific log type.

VALUE MANAGEMENT: A LOOK AHEAD

The value management concept is a systems process that will change the structure of the wood business. State-of-the-art mill equipment coupled with a sophisticated log allocation model will extract the optimum from the log.

The most successful firms will be capital intensive, with an evolving manufacturing process representative of the then-current state-of-the-art. The industry will become increasingly tree oriented rather than product oriented. The unintegrated, product-oriented producer will have an increasingly difficult time competing for a relatively scarce resource, the tree.

REFERENCES

Baldwin, R. F. 1981. *Plywood Manufacturing Practices*. Rev. 2d ed. San Francisco: Miller Freeman Publications.

Carroll, M. N. 1977. Utilization of Wood and Forest Residues in Reconstituted Panel Products. *Forest Utilization Symposium Proceedings, Winnipeg, Man., September 28-30, 1977*. SP No. 6. Forintek Canada Corp.

Darr, D. and Fahey, T. D. 1973. *Value for Small Diameter Stumpage Affected by Product Prices, Processing Equipment and Volume Measurement*. Abstract of Research Paper PNW-158. Portland, Oreg.: Pacific Northwest Forest and Range Experiment Station, USDA Forest Service.

Dobie, J. 1977. Sawmills. *Forest Utilization Symposium Proceedings, Winnipeg, Man., September 28-30, 1977*. SP No. 6. Forintek Canada Corp.

Donnelly, D. M., and Worth, H. E. 1981. *Potential for Producing Ponderosa Pine Plywood in the Black Hills*. Resource Bulletin RM-4. Fort Collins, Colo.: Rocky Mountain Forest and Range Experiment Station.

Fahey, T. D., and Hunt, D. L. 1972. *Lumber Recovery from Douglas-fir Thinnings at a Bandmill and Two Chipping Canters*. Research Paper PNW-131. Portland, Oreg.: Pacific Northwest Forest and Range Experiment Station, USDA Forest Service.

————. 1975. *Lumber Recovery from Grand Fir Thinnings at a Bandmill and Chipping Center*. Abstract of Research Paper PNW-186. Portland, Oreg.: Pacific Northwest Forest and Range Experiment Station, USDA Forest Service.

Kruppa, R. A. 1979. Getting the Most out of a Log. *Logging Management*, December.

Plank, E. 1979. Lumber Recovery from Live and Dead Lodgepole Pine in Southwestern Wyoming. Abstract of Research Paper PNW-344. Portland, Oreg.: Pacific Northwest Forest and Range Experiment Station, USDA Forest Service.

Pnevmaticos, Y. C., and Kerr, R. C. 1980. *Yield and Productivity in Processing Tree-Length Softwoods*. Technical Report 507E. Ottawa, Ont.: Forintek Canada Corp.

Snellgrove, T. A. 1975. Lumber Potential for Cull Logs in the Pacific Northwest. Abstract. *Forest Products Journal* 26(7):51.

Snellgrove, T. A.; Henley, J. W.; and Plank, M. E. 1975. *Lumber Recovery from Large, Highly Defective, Low Grade Coast Douglas-Fir*. Research Paper PNW-197. Portland, Oreg.: Pacific Northwest Forest and Range Experiment Station, USDA Forest Service.

Weyerhaeuser, G. 1965. *The Weyerhaeuser High Yield Forest*. Tacoma, Wash.: Weyerhaeuser, Co.

Nine
Capital and New-Technology Planning

Successful forest products companies have well-planned capital expenditure programs and capital-intensive facilities. The ongoing performance of each company, whether large or small, is strongly influenced by its capital spending decisions, decisions that affect long-term flexibility and earning power in a volatile and cyclic business.

"They [capital expenditures] are the engine which drives the balance sheet," commented a chief operating officer for a large integrated firm located in the southeastern USA. This manager recognizes that the required timberlands and long-term manufacturing facilities are huge investments that represent a commitment to the future.

For example, mill margins are being squeezed by rapidly escalating stumpage prices. In recent years stumpage prices have increased annually at a rate of 2% to 4% over inflation in the U.S. South and 6% to 10% in the West. Timely and well-conceived capital expenditures are required to offset these and other cost increases.

States a Pacific Northwest mill manager, "What we've been able to do with the new mill is salvage 8-foot and longer logs with a minimum 4-inch top, many of which were left to rot on the forest floor" (Griffin 1979, p.12). This dimension lumber mill utilizes second-growth Douglas fir and hemlock logs that may run 60 to 80 stems to the truckload.

This mill represents a recent capital investment and is typical of the newer mills, mills designed specifically to cope with a changing resource and a turbulent business environment. Risk and uncertainty are vital parts of the capital investment equation; each must be acknowledged and clarified.

CAPITAL PLANNING CONCEPTS

Forest industry projects are dominated by operational considerations such as improving utilization, increasing man-hour productivity, reducing costs or increasing market share. The capital planning process—also termed capital budgeting—must be simple, and it must be defined in a logical fashion. Senior management, financial personnel and the various staff groups along with the operating manager must be comfortable with the investment. Gaining understanding and obtaining the "biggest bang for the buck" involves assessing the businesses the company is in (or might get into) and the types of investment opportunities available in each.

The Panorama: Evaluating the Company's Businesses

The individual businesses—softwood lumber, structural panels and/or other manufactured products—are evaluated for desirability. This evaluation requires a detailed analysis of present and future prospects.

In this evaluation, or *panorama*, as it is sometimes called, each business is ranked as to industry attractiveness and the perceived strength of the individual companies competing in that business. For example, a company might conclude that producing reconstituted board in Minnesota is more attractive than making softwood lumber in the Intermountain West.

The evaluation will classify each business and the individual conversion units within that business as attractive growth situations or, on the other end of the scale, as "harvest" or "asset redeployment" situations. *Harvest* is a management buzz word that translates simply into "run it until it is no longer economical; only must-do capital projects will be considered in the interim." *Asset redeployment* simply means "sell it for what you can get; then invest your money somewhere else."

Capital expenditures and modern equipment—as in this automated lay-up line—are the engines that drive the balance sheet. (Courtesy *Forest Industries* magazine)

The Project Type: Sustaining or Generative?

Another capital planning concept recognizes two types of capital projects: sustaining and generative. The latter are sometimes called offensive investments, while the former are identified as defensive capital expenditures.

Generative (offensive) projects are defined as those that increase earnings from the existing base level. Generative projects decrease labor, increase income from sales, reduce waste or increase yield and/or reduce costs. An improved bandmill carriage setworks is one example of a discrete generative project in a sawmill; shutting down an obsolete lumber operation in favor of a new high-technology mill would be another example.

Sustaining (defensive) capital investments are defined as those that are intended to maintain and prevent deterioration of the current earnings level. A forklift replacement might be a department example; a fire equipment installation could be another. A pollution control system may be yet another plantwide defensive project. The implementation of a pollution control project may occur as the result of an operating mandate, or by the demand of an internal staff group or an external government agency.

A defensive project may sometimes yield a return, but that return is secondary to the primary goal of defending the current income stream while complying with business requirements established by the company, its vendors (such as an insurance company) or a government agency. Generally, defensive projects are must-do projects. Replacement or the scope of a rebuild job will depend on the serviceability of the equipment and the needs identified.

Business Growth: Three Views

A firm's statement of purpose (the mission defined more specifically) and its annual review and updating of the strategic plan provide the underpinnings for the capital planning process. These documents define the growth needs and the growth philosophy of the firm. Some companies may have sharply defined views; others rely on the entrepreneurial abilities of their individual business or location managers. Still other firms may combine aspects of some or all of the views or may subscribe to a view that is peculiar to the management style of an individual.

There are three prevailing views regarding growth and capital spending. These views can be summarized as follows:

- *Modernization and updating are necessary on a continuing basis to compete successfully for logs and timber sales.* This view is common among firms that purchase 60% and more of their logs or timber from outside sources at fixed prices.
- *Capital projects should enhance the value of the land and timber assets.* A successful project will both improve short-term results and increase the present value of the firm's land and timber assets. Weyerhaeuser is a good example of a company with this view.
- *Capital projects must maximize net income.* Firms that purchase timber at a price that fluctuates in response to finished-product prices—British Columbia firms, for example—share this view.

THE CAPITAL PLANNING PROCESS

Detailed capital planning is an outgrowth of the annual review and revision of the strategic plan. The long-run and short-term resource allocation decisions and the marketing strategies developed in the strategic planning process are necessary ingredients of the capital planning process. Capital planning is usually done in the spring prior to the calendar year in which the individual projects will be implemented. More complex projects may take several years of planning. The capital budgeting process falls into a logical sequence of activities.

Cash Forecast

A determination or accurate forecasting of the cash available from both internal and external sources over and above the needs of the ongoing business for implementing capital projects is a preliminary task. This provides a planning number or target for the planner. This cash forecast includes a definition of sources by type and cost, along with an understanding of the obligations that may be a requirement for obtaining the funds.

Gross Feasibility Analysis

The next planning activity is the gross feasibility analysis. This analysis identifies and ranks the generative (offensive) projects within a selected business segment.

Figure 9.1 lists the action steps used to identify and rank the generative projects within a business or an individual mill. A company closely adhering to a statement of purpose with a detailed overall capital investment objective will usually opt for fewer, larger, cost-effective projects in existing facilities or the complete prairie start of a new plant.

Action Step 1—Determine the Objective. The first step in a gross feasibility analysis is to determine and clearly state the objective of each capital project. Let's use a sawmill example to illustrate the process. The following are key variables in a sawmill operation:

Lumber recovery factor (LRF)
Lumber sales mix
Man-hour productivity
Volume and type of logs processed

Figure 9.1. Gross feasibility analysis action steps.

- Determine the objective.
- Identify leverage areas.
- List and evaluate potential opportunities.
- Analyze solutions.
- Calculate the capital costs.
- Evaluate and rank opportunities.

The project participants may decide that the overall objective for a (specific) capital project is to increase the yield of primary product 5% and more from the present log supply.

Action Step 2—Identify Leverage Areas. Leverage areas are places where capital improvement projects will provide a disproportionately large return for the dollars expended. The second step in a gross feasibility analysis is for the manager to identify areas where the stated objective (or more) is attainable. This activity may include obtaining the facts at the headrig, the edgers or the trimmer in the sawmill; it may include examining the milling, drying or board-making process in the OSB or waferboard plant.

Action Step 3—List and Evaluate Potential Opportunities. The manager next lists and evaluates all potential opportunities at each of the identified leverage areas. Figure 9.2, part 1 is an example of an evaluation for one opportunity at one leverage area in a sawmill.

The 9-ft bandmill with its poor size control and excessive variation in sizes was identified as an individual mill leverage point for increasing yield. Replacing the carriage setworks could reduce this variation. A study is made to evaluate this potential opportunity, and it is determined that the overall thickness variation can be reduced by 0.047 in. per sawline.

Action Step 4—Analyze Solutions. The proposed solution (capital project) is analyzed by determining the savings and costs associated with it. For the sawmill example, the savings calculation is shown in Figure 9.2, part 2: a forecasted first-year net benefit of $230,250. The benefit calculation uses the contribution approach; it identifies relevant incremental savings and operating costs under the base case (present operating costs, equipment and manufacturing practices) and the base case alternatives (the proposed capital project or projects).

"What difference will it make?" or "How will it appear on the statement?" are the two questions generally asked to identify the contribution for a specific project proposal. The savings calculation represents the benefits of the proposed capital project.

Action Step 5—Calculate the Capital Costs. The project cost is then identified. This is a rough-and-ready calculation and is useful only for the gross feasibility analysis. Purchased machinery will represent only a percentage of the total installed cost; the actual percentage will vary according to the equipment type.

Other cost considerations which must be taken into account in figuring the project cost are as follows:

- What engineering is required? Turnkey outside engineers and a master mechanic using a soapstone are the two extremes.

- Has the project been installed previously? By whom? What skill and familiarity will this provide?

- What is the design of the facility in which the project will be installed? Is it an older building with numerous support columns and a wood floor or a newer, more spacious facility?

- What commitment will there be on the part of the vendor? Will this commitment prevent a costly trial-and-error installation?

The resulting preliminary cost estimates are useful. Each provides the desired information with enough accuracy for a gross feasibility analysis. The preliminary estimates also provide the framework for detailed engineering, which will occur prior to completion of the formal authorization for expenditure (AFE) for any capital project.

Figure 9.2. Gross feasibility analysis savings calculations.

Part 1: Potential Yield Improvement from New Carriage Setworks

Statistics

Operating hr/yr, 2 shifts	4,040 hr
Average face width	2.35 ft
Average piece length	32.8 ft
Carriage pass/pc	1
Average number of rough lumber cants/min	2.11

Reduced actual cant thickness attainable

Sawing variation within each piece:	Present (1982) performance	0.089 in.
	Expected (test results)	0.042
	Net reduction	0.047 in.

Part 2: Calculation of Savings—Replace Carriage Setworks

Calculation of savings (benefits)

Based on a size reduction of 0.047 in. on each piece that is cut, the volume of fiber upgraded per year is:

$$W \times L \times T \times P \times 60 \times 4040 = ft^3$$

where: W = 2.35 ft average width
L = 32.8 ft average length
T = 0.047 in. reduced thickness variation
P = 2.11 pc/min
60 = min/operating hr
4,040 = hr/yr with 2 shifts

$$(2.35' \times 32.8' \times \frac{0.047"}{12}) \times 2.11 \times 60 \times 4040 = 154,409 \ ft^3$$

Calculation of increased value available

Critical assumptions: 60% lumber
40% chips
100%

Product	Ft³	Conversion factor	Product volume	Incremental value/unit	Total
Lumber	92,645	14.7 Bf/cf	1,362 MBF	$157.46	$214,461
Chips	61,764	.013 BDT/cf	803 BDT	73.40	58,940
	154,409				$273,401
Less shavings loss:					
	154,409	.013 BDT/cf	2,007 BDT	21.50	43,151
Net savings per operating year					**$230,250**

Action Step 6—Evaluate and Rank Opportunities. After all the potential projects are identified, the task then becomes one of evaluation and ranking.

A *discounted cash flow analysis* is useful for initial project scoping and evaluation; it is usually used with projects that have lives of 10 years or more and for then evaluating timberlands. This analysis requires several critical estimates, such as:

- The amount of the expected investment
- The amount and timing of the necessary cash flow
- The hurdle rate (the minimum rate of return that could be expected if the capital projects were not approved and the funds were directed elsewhere)

For a discounted cash flow analysis, the capital cost (including associated installation and startup costs) is calculated as a present value. The tax-shield effect of depreciation along with the expected tax rate is used in this calculation. Inflation is considered when calculating net savings.

The net savings are forecasted for each year over the expected life of the project, with each year discounted to the present value. The sum of these savings then represents the total *net present value* (NPV).

The present value of the net saving is then divided by the applicable project capital costs and expenses to arrive at a ratio. This ratio is a useful tool for project evaluation. A further refinement is to factor in the hurdle rate that the company must have to justify an investment. The presumed cash flows are discounted by this amount to arrive at the excess return.

The *payback analysis* is another effective evaluation method. This method is also called payout or payoff; it is an easy-to-use evaluation tool. It measures the time it will take to recoup the original dollars invested based on expected cash flow. The payback analysis is a useful tool when:

- The precision of the estimate is not crucial and a preliminary evaluation of a number of projects is necessary
- A weak cash and credit position has an important bearing on the decision
- The proposed project offers a substantial degree of risk over the long term

The payback analysis is usually present on the formal AFE when this document is presented to the capital committee or the board of directors for approval.

The *analysis of risk* is a key part of the evaluation and ranking process. Risk analysis measures and evaluates the likelihood of achieving the expected results. External factors such as market demand or price forecasts are considered. Manufacturing costs such as logs and labor are also considered. Internal factors such as people and equipment are evaluated too. The track record of the operating manager is evaluated on the basis of past and expected management performance in achieving results.

A recent study group surveyed a number of forest products companies to determine if and how risk was handled or factored into the capital budgeting process.

Of the responding companies, 44% indicated that the cost of capital was increased or the project life or its benefits were adjusted downward. The majority used some type of sensitivity analysis to account for risk.

Sensitivity analysis is a way of systematically working through the effects of assumed changes in revenues, benefits and costs on the overall return. Most often the major projects are evaluated at three levels of assumptions: the expected, the optimistic and the pessimistic.

Use of these three assumption levels is a relatively unsophisticated method, although quite effective. Other more sophisticated methods are well adapted to computer programming. Each can be as detailed and complex as the resourcefulness of the manager, the skills of the programmer and the capacity of the computer. These methods include certainty equivalents and risk-adjusted discount rates.

Sophisticated computer models make it possible to estimate the key variables for the major projects and the range of possible outcomes for a score or more of key variables.

As investment analysis tools, evaluation techniques and computer models are not substitutes for judgment; they merely serve to guide its application. A new West Coast veneer plant is an example.

The project had been studied in detail, with all possible mill and market variables considered. The mill started up as a state-of-the-art manufacturing facility; logs were plentiful and a market for green and dry veneer was readily available. What was not readily available to the planners was the knowledge that the union local would not support the startup of a third, or graveyard, shift.

"We had too much union troubles running the third shift to continue," commented the manager. Yet the third shift was needed to amortize a greater product volume over the capital costs associated with the new technology employed.

This company had difficulty picking a winner on this project because the labor situation could not be quantified. Picking a winner is something more than a gross feasibility analysis and the risk analysis that is part of the process. Picking a winner also involves the following:

- Measuring the risk involved both subjectively and objectively; outlining the critical assumptions and how each will be controlled.
- Avoiding the common pitfalls (Figure 9.3). Determining how the savings or benefits will appear on the statement.
- Sidestepping the fads and fancies; picking the tenth copy rather than the first. The untested fad or fancy should be developed by a company that can amortize the initial costs over a number of mills.

The Authorization for Expenditure Document

The authorization for expenditure (AFE) is the formal document that is prepared as a result of the gross feasibility analysis and evaluation. Sustaining investments that are required for continued operation also use an AFE format.

Content of the AFE. The AFE document identifies the capital project, its scope and the expected costs and results. The actual document may be simple or complex; the degree of complexity is usually proportional to the size and diversity of

Pitfall	Description
Should have worked	A project not adequately engineered and evaluated before installation
The idea	Custom-made first-of-its-kind equipment
Now you see it, now you don't	A cost savings project offset by cost transfers to other departments
Head in the sand	Equipment purchased to solve apparent rather than real problems
New wine in old bottles	New equipment tied to obsolete equipment and systems
The ego investment	Equipment purchased as a want rather than a need
The Titanic	A grandiose scheme or combination of losers so large that the business entity sinks while the crew bails futilely

Figure 9.3 Pitfalls of capital project planning. (Source: Modified from Smith 1974, p. 51)

the company. Figure 9.4 lists the titles used for such documents by a number of forest product companies.

As the titles imply, the actual document may act as both the approved justification and the spending authority, or it may be an appropriation request, with actual release of funds subject to further review. The latter approach allows a steady flow of funds. Planning is done with greater lead time; usually there is a backlog of prioritized capital projects that can be immediately implemented when funds are available.

Figure 9.5 is an AFE checklist. Note that the content is quite similar to the gross feasibility analysis but is more detailed. Savings estimates are refined; sufficient engineering has been completed to cost out the project within an allowable overrun or underrun.

Project management considerations, such as installation schedules and cost control methods, are identified. Three questions should be asked:

Figure 9.4 Various titles for authorization for expenditure document.

Company	Title
Boise Cascade	Authorization for expenditure (AFE)
St. Regis	Job order purchasings (JOBS)
Weyerhaeuser	Appropriation request (AR)
Simpson	Approved cash expenditure (ACE)
Champion Intl.	Appropriation request authorization (ARA)
Georgia-Pacific	
Capital good	Authorization for expenditure
Mobile equipment	Authorization for mobile equipment

- Are the savings estimates realistic? How will savings show in the operating statement?
- Has the engineering been sufficiently completed?
- Are target dates realistic?
- Is the CPM/PERT schedule (see Chapter 6) completed?
- Have all the startup costs been considered?
- Are estimates current at the time of approval and funds released?
- What provision has been made for cost control during the project?
- What controls and methods will be used to ensure that forecasted results will be achieved?

Figure 9.5 Authorization for expenditure checklist.

- Is the information accurate enough that I would place my money in the investment?
- Can I determine exactly how this project is to be implemented?
- Will others who read this document and the associated analysis obtain a clear understanding of the project and its benefits?

Investment Analysis. Answering these questions also requires a more refined investment analysis. This investment analysis usually involves the discounting of future cash flows to establish the *net present value* (NPV), as described earlier, or an *internal rate of return* (IRR). IRR, a discounted cash flow method, is defined as the interest rate that discounts the present value of the future cash inflows from the project with the cost of the investment.

Cited a writer, "On a conceptual basis, these measures have been conclusively demonstrated to be superior to such other frequently encountered capital management criteria as payback and average rate of return" (Bailes, Nielsen and Wendell 1978, p. 1).

These measures are intended to answer two questions:

- Should the company accept the proposed course of action as outlined in the document?
- Given competition for funds, which AFEs should be chosen?

Another indicator is useful also; it is the *return on capital employed* (ROCE). This indicator identifies the expected yield on the original investment by relating average annual after-tax net profit to total capital employed (both fixed assets and working capital). The ROCE is an indicator of what can be expected in overall company return.

Selection of the indicator to be used in the investment analysis will be determined by the complexity of the investment and by the intricacies of the financial function and its acceptance by top management and the board. Whatever the method chosen, it is important that the people involved do not put on their rose-colored glasses but ascertain that the project merits approval and implementation.

Post-decision Procedures

Post-decision control procedures are centered on two areas: the monthly capital expenditure report and the post-completion audit. The *monthly capital expenditure report* tracks the actual engineering, purchasing and installation costs by project item. This control report should be sufficiently detailed to enable the project manager to track actual results against the plan results as defined in the AFE.

The *post-completion audit* then tracks the actual benefits achieved versus the plan document. The benefits identified are evaluated along with the other project cost-and-benefit facts. The result is a detailed audit to determine if the project goals were achieved.

The post-completion audit is a useful tool. It provides feedback to the project participants and to top management. This in turn assists the manager and the planners to prepare more accurate documents for future projects.

These post-decision procedures—the monthly capital expenditure report and the post-completion audit—combine with the gross feasibility analysis and the authorization for expenditure to provide a capital expenditure management system that is responsive to change. Aggressively dealing with change is at the core of a profitable forest products operation.

REFERENCES

Bailes, J. C.; Nielsen, J. F.; and Wendell, S. 1978. *Capital Budgeting Practices in the Forest Products Industry*. Monograph. Corvallis, Oreg.: School of Business, Oregon State University.

Dunn, R. A. 1981. *Management Science—A Practical Approach to Decision Making*. New York: Macmillan.

Edwards, J. D., and Black, H. A. 1976. *The Modern Accountant's Handbook*. Homewood, Ill.: Dow Jones–Irwin.

Griffin, G. 1979. Mill No. 5 Initiates Small Log Program. *Timber Processing Industry* 4(10):10.

Helfert, E. A. 1972. *Techniques of Financial Analysis*. Homewood, Ill.: Dow Jones–Irwin.

Horngren, C. T. 1967. *Cost Accounting: A Managerial Emphasis*. 2nd ed. Englewood Cliffs, N.J.: Prentice-Hall.

Smith, J. A. 1974. Watch Out for These Common Capital Expenditure Pitfalls. *Forest Industries* 101(1):50 (January).

Section Four
THE PRODUCTION PROCESS

There is a tendency for a business to become slack and easygoing, particularly in good times. The manager's challenge is to overcome the inertia and concentrate resources as if each day is the bottom of the market.

CHAPTER TEN

Ten
Managing the Fundamentals

The forest products industry has experienced six roller coaster–style business cycles since World War II. Each has been characterized by exhilarating, high lumber and panel prices, with demand frequently outstripping the capability to produce. These highs have been followed by lows: gut-crushing lows during which products were overly abundant and prices were well below the cost of manufacture.

An industry newsletter quoted a manager's comment during the downside of a recent cycle, "If we had enough money, we'd hire Don Meredith to sing: 'Turn out the lights, the party's over . . .' " (Southern Forest Products Assn. 1981).

One entrepreneur thought he had a better idea than singing: get into business again. After going broke on a market low, he simply organized a new business on the following upturn and called it TOMCO, for "Try Once More COmpany." He weathered the next downturn and regained more confidence as the market improved. This renewed confidence was translated into new, sophisticated hardware and expensive timber purchases. He went broke once again. He didn't feel much like singing when he lamented as if to music, "I have no hide left" (Albany *Democrat Herald* 1982).

It's a tough and turbulent business; the survivors are those who manage the fundamentals well. The key to managing the fundamentals is the outlook of the decision makers and how well each understands and executes the time-tested fundamentals, or basics, as they are sometimes called.

IDENTIFYING THE FUNDAMENTALS

Each business cycle has been followed in turn by a changing business environment as the national business economy has evolved into a world economy. This world economy is characterized by increased regulation, more structured social change and an evolving demand pattern. Each adds to the other to obscure the manager's outlook.

"We're looking at a much different future. . . . None of us knows quite how the whole thing will come back together" stated a timberlands manager (Briggs 1982,

p. 136), during a recent downturn, as his company plus other forest products producers sought to survive and cope with the changing business environment.

The future will be determined by the perceptions of the manager and his skillful use of the fundamentals.

A Questioning Attitude

A questioning attitude is a management fundamental, a fundamental that is a prerequisite for managed change. The manager can never assume that tomorrow will necessarily be an extension of the present. And there is nothing like a huge loss to bring the importance of a questioning attitude into sharp focus. The $320 million write-off of ITT Rayonier's Port Cartier, Quebec, project is one such example.

"The forested land base was gigantic; it was nearly the size of the state of Tennessee. But there were so many questions we should have asked prior to committing resources," commented a manager not directly associated with the initial project. His subsequent statement sums up the unasked question: "If there is a whole big patch of trees left to cut, there is generally a . . . good reason!"

Identifying the "good reason" is usually the end result of asking questions like these:

- If it hasn't been done, there must be reasons. What are they?
- I heard it, but I am not certain I believe it. What are the facts?
- If everyone is doing it, why are we still doing it? Or the reciprocal: If we are doing it and no one else is, what do we know that they don't know?
- Why can't we do more or do it better or faster?
- It's running well. What is about to go wrong?

Lack of a questioning attitude soon leads to obsolescence and failure. Both are closely tied to a refusal to accept change. The issues of change will come in different sizes and shapes in the years ahead. The successful manager will recognize a questioning attitude and a quest for facts as management fundamentals.

Accountability and a Can-Do Attitude

It is fundamental that the manager establish credibility by demonstrating his ability to deliver on targeted goals in production, quality, cost control, man-hour efficiency, reports, forecasts, plans and other activities. His task is to allocate resources, assign personnel and provide direction for his organization.

There is a tendency for a business to become slack and easygoing, particularly in good times. The manager's challenge is to overcome the inertia and concentrate resources as if each day is the bottom of the market. The manager's can-do attitude will effect results through others. He must be prepared to take the slack out of the organization if that's what it takes to get the job done.

One such manager is an example. "On walking inspections of the facilities, he stops to chat with workers and takes note of everything. 'If there are cigarette butts on the floor, you can bet it's a badly managed plant.' He says his management style involves 'testicle escrow.' He explains: 'We're all good friends, but our managers know they have to perform. I like to say they have one testicle on deposit' " (Solomon 1978, p. 130).

This can-do attitude spills over into the follow-up of assignments. Commented a 30-year-plus veteran plywood manager, "Don't give any instructions that you can't/won't follow up on. . . . You need to put as many numbers as possible on all instructions."

Accountability and achievement are tied directly to assignment control. Assignment control requires, first, that a manager know the strengths of people. Particularly important is identifying people who have a proven record of performance. What do they do well? Where do they belong?

Secondly, people should be assigned to jobs in which the application of their strengths can produce results. This requires that each individual be assigned the opportunities that are right for him or her and the needs of the firm.

Keeping Score

A questioning attitude charts a course of action; a can-do attitude leads and directs that action. Keeping score is a management fundamental that tracks the results achieved.

"The operation of the mill revolves itself around the strategic plan, stated written objectives, programmed schedules and completion dates, with resources budgeted to the plan. A plan is no better than the controls it has to determine if the plan is working. Control is performance standards, measurement, evaluation and correction," commented a senior manager during an interview.

Daily reports, interim statements and monthly profit and loss results are all vehicles for keeping score. Increasingly, periodic statistical reports are available from computerized process control functions. Sophisticated process control hardware with software reporting and analysis features provides accurate periodic reports on work in progress or results achieved.

These scorecards should be evaluated on the basis of attainable standards. The attainable standard should correlate closely with the monthly statement. "If it doesn't appear in the statement, it never happened. . . . If a question arises as to what basis production numbers, quantities and other statistical information are to be collected, . . . the same numbers in the same format should appear in the statement," further stated the same senior manager.

Mill Management Principles as Fundamentals

There are a number of time-tested principles that over the years have developed into a fundamental part of management. The ABC Classification of Costs is one of these.

The *ABC Classification of Costs*, or Pareto's Principle, as it is sometimes called, is based on the premise that about 80% of the costs are represented by 20% of the items.

The first and most important step is to find out where the costs are. These data are usually fairly easy to obtain; then they must be classified to make the best use of the information. The manager spends his time working on those costs that have the greatest cost-reduction possibilities. Initially, this concept will highlight two prime manufacturing costs, wood and labor.

The *Leverage Principle* is a cousin to the ABC Principle. The Leverage Principle states that each cost center has at least one leverage point: a point at which a modest effort will reap large benefits. The leverage point that represents the greatest benefits should have the highest priority.

For example, an operator determines that reducing the decimal thickness of the veneer will provide a yield increase of 2% while maintaining an acceptable panel tolerance. An additional 2% may also be available with a new clipper control. While both projects are worthwhile, the first would have the highest priority because it requires no capital expenditure.

Commented a senior vice-president of a large company, "With raw materials from 65 to 70 percent of the cost of lumber production, recovery of finished product per unit of raw material, along with reduction of labor cost per unit, is a very important economic leverage point."

The speaker went on to describe the effect of using the twin principles, ABC Classification of Costs and the Leverage Principle: "For the last five years, our nine southern plywood plants have occupied eight of the top ten positions of 21 plants that reported to the American Plywood Association in terms of either margins or productivity" (Bingham 1981).

The *Little/Big Things Concept* is a third management principle. This principle is based on the assumption that if the small things are taken care of, then the big things will take care of themselves. The following is an instance of this principle at work in a manufacturing plant.

A manager noted during his rounds that a shop-grade plywood panel, rather than a less expensive cull board, was being used as boxcar blocking. Closer examination identified incorrect bulkhead blocking, which could have resulted in a substantial shipping damage claim.

The car was unloaded and individual units of plywood were check-graded at random. Excessive undergrade was noted in each. The rest of the department was then inspected, with further deficiencies noted. The finishing department supervisor of this large softwood plywood plant was new to the job. He in turn had supervisors who didn't pay sufficient attention to the work in progress. The manager followed the chain of relatively small deficiencies through the department and resolved each before they developed into a later stream of serious problems.

The Little/Big Things Principle works well on paperwork such as performance reports and statements. One relatively unimportant number may lead to a question; the question then provides the vehicle for further in-depth investigation. The investigation often directs the manager to more serious problems in the making.

The three principles mentioned are a few of the many. They represent methods for controlling the fundamentals. Let's observe the result of their effective use.

INDICATORS OF WELL-MANAGED FUNDAMENTALS

During a recent 2-year period the author visited more than 50 individual lumber and panel manufacturers in North America. The initial observations were confirmed during later visits, including a tour of like facilities in Scandinavia. Detailed observations were made; quite often the notes gleaned from a trip were sufficient to construct a pro forma; a standard format permitted comparisons.

From those observations it's possible to characterize the outward indicators of well-managed fundamentals as follows:

- Each operation paid close attention to manufacturing techniques and methods, production planning and individual control over quality.
- Tight accounting and work-in-progress methods were in evidence. Loads were well stacked and identified at each transfer point.
- In-process inventories were as low as practical. The better performers had a below-average volume of in-process or finished goods for their particular manufacturing process.
- A work plan was organized for results; jobs were defined and their accomplishment was measured by demanding but achievable standards. The performance to those standards was communicated to each team member.
- Each business entity had a definable goal with which individual employees could identify. The employees were usually knowledgeable and knew the importance of their individual contributions.
- Open relationships were developed that reflected the worth of the individual employee with little regard for racial or ethnic background.
- Employee attitude mirrored the condition of the physical facility and the conduct of the staff. Good housekeeping in the workplace, office, restrooms and lunchrooms was a standard. The employees were responsive to questions and smiled and conversed in a businesslike manner.
- Oral and written communication reflected management's confidence in each team member.
- An ongoing commitment was made to provide formal and informal training and self-development opportunities.

During further conversation with the key managers, it was apparent that an intimate knowledge of the manufacturing process is used as a competitive edge on the competition. High-technology changes are sought when sufficient improvement can't be accomplished through management control of the fundamentals. Spending is tightly controlled: a dollar is spent only when considerably more than a dollar will be returned.

Lastly, the most important visible characteristic of operations that manage the fundamentals well is the fact that each of the companies has survived and prospered through more than one cyclic downturn with fewer than average layoffs and curtailments. That's the value of closely controlled fundamentals in the forest products industry: survival through the lean times and better-than-most profitability during the periodic market upturns.

REFERENCES

Albany Democrat Herald. 1982. TOMCO Boss: 'I Have No Hide Left.' June 29.
Bingham, C. W. (senior vice president, Weyerhaeuser Co.). 1981. Weyerhaeuser.

Presentation to the Atlanta Society of Security Analysts, Atlanta Ga., December 2.

Briggs, J. A. 1982. Forest Products. *Forbes* (1):136–139.

Drucker, P. F. 1980. *Managing in Turbulent Times.* New York: Harper & Row.

Loomis, C. J. 1979. How I.T.T. Got Lost in a Big Bad Forest. *Fortune*, December 17, pp. 42–55.

Solomon, S. 1978. The Tough Prof at Georgia-Pacific. *Fortune* (12):128–130.

Southern Forest Products Assn. October 17, 1981. *SF Lumber Crisis Hotline.* New Orleans, La.

Eleven
Log Processing

Log processing activities commence when the tree is on the ground and conclude when the log segment is introduced into the conversion process. These activities are more than just getting the log to the mill; each is expected to be cost effective and yet minimize subsequent processing costs. Figure 11.1 describes key activities.

LOG PROCESSING ACTIVITIES

The first activity, *felling*, is not actually considered log processing, although it is the foundation for the sequence of activities that follows. Improper felling will cause breakage and introduce other mechanical defects, such as shake. The result-

Figure 11.1. Key log processing activities: tree to mill.

Activity	Description
Felling	The standing tree (stem) is severed from the stump.
Limbing	The limbs are trimmed flush with the bole of the tree.
Bucking and topping	The log or stem is cut into smaller segments; the unmerchantable top portion is severed from the stem.
Scaling	The log or stem is graded and measured for volume.
Storage	The logs or stems are gathered and stored as an intermediate step prior to transportation and/or processing.
Sorting	The stems, logs or log segments are physically separated into like groups by attributes such as size, quality, length and/or species.
Hauling and Transfer	The stems, logs or log segments are conveyed from one point in the process to another.
Debarking	The bark is removed from the log or log segment prior to conversion into a forest product.

ing defects may not show up until the log is in the lathe or on the headrig. In addition, improper felling makes limbing, bucking and topping more costly and difficult. The difficulties encountered often translate into additional breakage and log damage.

Limbing is important to the subsequent log processing operations also. Improperly trimmed limbs, those that are not trimmed flush with the bole, can result in conveyor jam-ups, broken conveyor chain and damage to the debarker. In addition, high-speed infeed systems for canters and chipping canters and sharp-chain infeed units require close control of the log as it is being profiled. Wasted peeling time

Bucking and topping are prime value-making decisions.

and lathe knife damage are problems encountered with untrimmed limbs and stubs in the plywood plant.

Bucking and topping, additional log processing activities, are prime value-making decisions. Choices of the "make it or lose it" type are made here. The goal is to maximize the net return when the resulting log is sold or converted into products.

Scaling, another log processing activity, was once an arduous art. This art is being reduced to a computerized measurement task as log rules evolve into volume formulas and the log itself becomes smaller and more uniform. The task of hand-scaling each piece is being replaced by sample scaling and weight scaling or by electronic scaling at the mill. Increased accuracy and more objective measurement criteria are the results of this evolution from the traditional forms of stick scaling to the new forms.

The *storage* and *sorting* functions become increasingly important as the industry shifts from water storage of several species or log types to dry land sorting and storage of a large number of species and log types—and a resulting higher piece count per unit of volume.

The other key activities—*hauling, transfer* and *debarking*—are all important parts of the process and are included in the stream of activities.

LOG PROCESSING PRINCIPLES

Figure 11.2 identifies the "vital few" from the many principles that are developing as log processing concepts evolve. These can also be termed monetary considerations because they have a direct effect on the bottom line. The six principles listed in the figure can be summarized as follows:

- Decision making is being automated and standardized by electronic and mechanical means.
- Log utilization decisions are best made close to the converting unit.
- Log processing steps and the complexity of those steps should be balanced closely with the costs incurred and the value obtained.

Figure 11.2. Log processing principles.

- Log processing decisions improve in direct proportion to a decrease in the distance to the final conversion process.
- Quantify human judgment whenever possible. Use electronic or mechanical means as tools to standardize and optimize decision making.
- High visibility and a close inspection of the stem or log segment results in improved utilization decisions.
- Log processing costs increase in direct proportion to the number of log processing steps and the number of choices available at each step.
- Log processing maintenance and operating costs increase in direct relation to the complexity of the hardware and rolling stock used.
- The greater the number of activity steps, standing tree to mill, the greater the fiber loss and the lower the resulting product yield.

NONMONETARY CONSIDERATIONS IN LOG PROCESSING

Nonmonetary considerations also weigh heavily in the activity steps (Figure 11.1) selected and the principles (Figure 11.2) observed. Typical nonmonetary considerations are as follows.

The Environment

Water transportation and storage are being phased out as citizens' concerns are translated into environmental regulations. These regulations will eventually rule out this storage and transportation mode.

Dry land hauling and dry land storage are not without concerns to the producer. Stringent environmental requirements are being placed on log-yard-runoff water quality.

Woods Crews

Hard work, long hours, uncertain work schedules and an urban-oriented society are taking their toll of loggers. It is increasingly difficult to recruit and retain quality woods help in adequate numbers. Activities previously done in the woods are being shifted to mechanized downstream machine centers.

Timber Sales Requirements

Federal and state forest services plus the Bureau of Land Management and other timber-holding agencies are increasingly dictating what to cut within the timber sale, how to cut and what to take from the woods. These regulations are forcing more and more marginal material into the mill yard.

EVALUATION IN LOG PROCESSING METHODS

The monetary and nonmonetary considerations plus a changing tree resource are the factors that are producing a three-stage evolution in log processing modes from woods centered to dry-land centered to merchandiser centered.

In the *woods-centered mode*, prebucked, presorted and scaled log segments are delivered to the mill or storage yard; water, the cheapest method of transportation, is used whenever possible.

In the *dry land sorting and storage–centered mode*, sorting, scaling and some bucking for value shift to a dry land sort yard. Transportation to and from the yard is by truck, water and/or rail. The dry land sorting and storage area may or may not be adjacent to the mill.

The *log processor/merchandiser–centered mode* usually includes a dry land storage area that has some sorting capability. Treelength or long-length material is gathered in the woods and then transported to the site. In some instances (as is the practice of Weyerhaeuser's Jacksonville, North Carolina, complex), the limbs are left on the unmerchantable top portion of the stem.

The woods task becomes little more than felling, bunching and hauling. A minimum of bucking is done, and that is limited to what is necessary to meet the hauling and merchandiser length requirements. Species with a relatively short form, such as loblolly pine, will require little activity, while a taller Douglas fir tree will need extensive bucking in the woods.

This evaluation in log processing is geared closely to the industry trend to multiproduct, integrated operations. This also ties closely to the resource evolution underway, as harvesting and processing of large trees is being phased out in favor of the smaller stem of the second and third forest. Because these trees are more uniform in size, they are readily adaptable to mechanized and automated handling. Human decisions can be strengthened with sophisticated electronics and computers. Decisions can be preprogrammed, stored in memory and triggered when the log is scanned.

Each solution, particularly the bucking decision, is geared to the requirements of the converting mill. Figure 11.3 identifies the log lengths predominately used in the various downstream converting facilities. These lengths range from 5-ft pulpwood to 40-ft and longer segments for pole and piling operations.

LOG PROCESSING SYSTEM SELECTION

Selection of a log processing system hinges on the results of long-term resource and product planning, as described in Chapter 4. This planning should forecast, with confidence, the sequence of harvesting for 5, 10, 15 and up to 30 years of the harvest cycle. The equipment and methods selected should correlate closely with this forecast.

Industrial engineering methods, particularly graphic methods such as a flow process chart, will assist the operator to organize and refine the transportation function, inspection, delays, and the storage and processing activity steps that occur during the process. The relevant information should include an analysis of ac-

Figure 11.3. Primary softwood conversion options.

Product	Log length predominately used
Lumber	• Single length • Random length in 2-ft increments from a specified minimum (usually 8 ft) to a maximum desired length
Veneer or plywood	• Peeler lengths in 8½-ft multiples • 8½-ft blocks
Poles or pilings	• Tree length trimmed to pole or piling lengths
Pulpwood or fiber	• Short length (5 ft or less) • Tree length to a specified maximum
Reconstituted board furnish	• Short lengths
Special products/outside sales	• Not applicable

The wheeled loader is the most popular type of handling equipment for unloading and transporting logs.

tivities to be accomplished, the time required for each activity and the distance and method of movement for the raw material. Simulation models and other computer aids (described in Chapter 6) are useful in the planning activity.

Choosing Handling and Storage Equipment

The operator has a number of options when choosing a handling and storage system. The most popular handling approach is use of a wheeled loader to unload and transport logs; cold-decking of logs is the most frequently used storage method. The various crane systems are also gaining in popularity.

Wheeled Loaders. Dry land log handling, as we know it today, began with the introduction of the Marathon LeTourneau wheeled loader in 1952. This unit, and the units that were subsequently developed (such as the Raygo-Wagner and the Dart), revolutionized the log yard handling function. A properly sized unit, usually of 80,000-lb lifting capacity and above, can unload a truck in one bit or a rail car in two. This equipment can build cold decks and transport logs to a sorter or another station, including the processor or mill infeed. Stacking height is one limitation of these machines. A crane can build decks up to 40 or 50 ft in height; a wheeled loader is limited to 12 to 15 ft without the aid of a stationary loader.

The wheeled loader combined with one or more stationary, hydraulically operated loaders is the most versatile handling, sorting and storage method. The smaller log loaders, such as the Caterpillar 966, can do limited sorting. This machine and other similar machines can preclassify incoming logs.

The hydraulic loader then sorts from the hot, or preclassified, pile; individual logs are placed into bunks, bins or pockets located around the perimeter of a circle. Each sort pile is positioned within reach of the shovel boom.

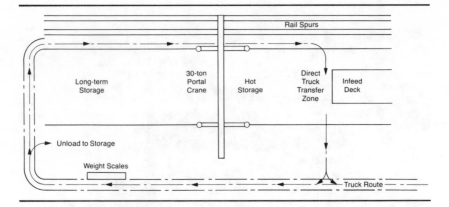

Figure 11.4. Portal crane: yard and storage layout. (Source: Hampton 1981, p. 190)

Cranes. Log-handling cranes come in all sizes and configurations. These can be classified into two main types: (1) those traveling on powered legs over rails and (2) stationary models with a boom that rotates through a fixed circle with a stationary center support. Both types use a cable-hoisted rotating grapple.

The *traveling crane* has been used in planer sheds for decades. It is also applicable to the log handling task, as the unit at the Plymouth, North Carolina, operation of Weyerhaeuser Co. illustrates. This 30-ton-capacity unit stands 104 ft high, with a clear span of 160 ft between the end supports and the parallel rails over which it travels. Figure 11.4 is a schematic of the layout for this crane.

The main girder is 240 ft long, which allows a 40-ft span past each leg or support. This crane is also described as a *portal crane*; it has the ability to pass hoisted loads from the cantilever sections of the crane girder through the traveling support legs.

Some *stationary models* also have this portal feature, although *jib boom cranes* such as the Kockums 90 or the LeTourneau JC 40 do not have this feature. These cranes have a counterbalanced boom that requires no outside support and therefore has no need for the portal feature. Figure 11.5 illustrates this crane type. The illustration shows a typical arrangement for unloading trucks, sorting logs and feeding one or more infeed decks.

This LeTourneau JC 40 model has a 40,000-lb-capacity grapple life and can unload incoming trucks at a rate of 8 to 12 per hour with 2 to 3 bits per truckload. Its 100 to 120-ft boom can build two concentric log storage decks, which fold into storage when the butts are oriented on the outer ring of the two circles. Note the contour of the surface area around the boom; it is intended to assist the operator as he forms the storage decks.

This stationary crane type is gaining wider acceptance in the U.S. Southeast, and it can be expected to see service in the other geographic areas as the timber profile continues the shift to smaller second and third-growth material.

Another crane type is the *log boom style*, which is seen frequently. This unit, often custom designed, is still fabricated and sold by a number of manufacturers

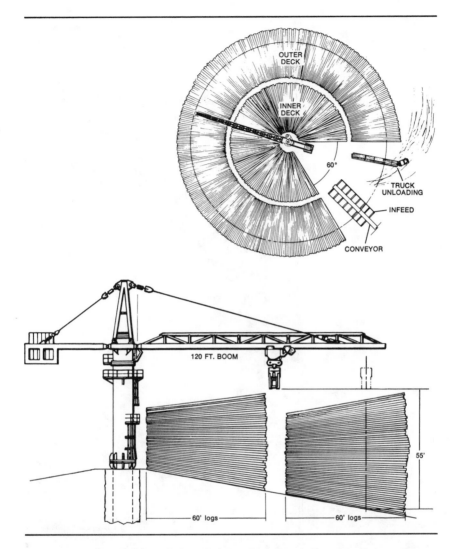

Figure 11.5. Typical jib crane layout. (Courtesy Marathon LeTourneau Co.)

and is installed from time to time. There are also a number of other configurations that are designed for a specific application. Most crane manufacturers have a number of models or variations available.

The crane becomes the workhorse for the application selected. Generally the initial cost for the crane system is higher than that for a wheeled loader system; the reduced surface area required for log storage plus favorable operating costs will weigh heavily in the selection of a crane system over the more common wheeled-loader/cold-deck handling and storage method.

The LeTourneau JC 40 jib boom crane is specifically designed for log handling.

Choosing a Log Processing/Merchandising System

Two considerations, the extent of the sorting done at the yard and the multiplicity of the conversion options downstream, will determine what is needed in log processing/merchandising hardware. Figure 11.6 describes and classifies the various systems.

Cut-to-Length Systems. In a cut-to-length (CTL) processing system, logs are cut to mill length in the woods or at another machine center, as is the practice at the chip mill illustrated in Figure 11.7. This chip mill configuration is fed by a jib boom crane onto a slasher deck. The merchantable portion of a larger log is retrieved as a measured and CTL segment. The resulting lengths are usually two-block multiples for the peeler plant or 16s for the sawmill.

Traditionally the woods crew prepared the CTL segments for the sawmill. The sawmill then debarked the CTL segments and ran them directly into a primary breakdown unit such as a canter or a chipping center. The setworks were changed frequently in response to the differing diameters. The Scandinavians have changed

Figure 11.6. Log processing systems.

Type of system	Millsite activities
Cut to length (CTL)	• Storage • Debarking • Scaling • Sorting (size, quality and species)
Log length Single conversion option	• Storage • Sorting (size, quality and species) if necessary • Bucking into segments
Multiple conversion choices	• Storage • Sorting (size, quality and species) • Bucking into segments • Allocation to appropriate conversion process
Tree length Single conversion option	• Storage • Sorting (diameter, length, quality and species) • Bucking into segments (tag ends chipped, sold or transferred)
Two or more conversion options	• Storage • Sorting (diameter, length, quality and species) • Bucking into segments using documented bucking criteria

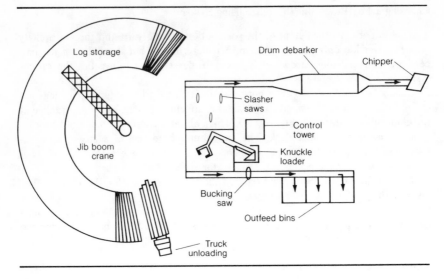

Figure 11.7. Chip mill with cut-to-length (CTL) retrieved.

all that. The CTL segments are still prepared in the woods but are sorted by species, quality characteristics and diameter when the logs reach the mill. The Tampella Oy sawmill at Porvoo, Finland, is an example.

There, CTL logs are unloaded from the trucks and placed on a live deck infeeding an automated barking and sorting processor. A Hammaras (Swedish) infeed deck assembly singulates the logs into one of the two debarker lines. A remote-controlled articulated grapple untangles the infeeding logs as necessary.

Downstream from the debarker the logs are scanned and are run through a metal detector. Each is then sorted into one of 34 bins based on scanner information. There are 15 bins for quality, 17 for size and 1 for wood containing tramp metal. The remaining bin is used for unmerchantable material.

The system evaluates the log for value and segregates the material by size so that fixed sets can be run for a particular log quality. This is much different from the variable sets common in North American mills. The result is close size tolerances that need no planing to satisfy European customers. Eliminating setworks delays between sets also increases sawmill productivity.

This concept is gaining wider acceptance, even in the North American industry. Commented an operator in Washington State: "By having a tight diameter sort we could run logs end-to-end with minimal set changes in the mill and increase our output tremendously. . . . By tight diameter sorting, you don't worry about machinery setting time. You just set your targets and let the machinery run" (Griffin 1982, p. 12).

Log-Length Processors. The most commonly used log processing system at both sawmills and peeler plants is log-length processing. This simple system bucks precut logs into predetermined lengths, such as plywood blocks or the most desirable lengths for the sawmill.

The sawlog system features a bucking station that will buck out the portion of the log suitable for the multiple sawmill option, such as a bandmill and a chipping center. Other systems include kickout pockets or transfer stations for other product types, such as plywood blocks.

Plywood mults are simply bucked into 8½-ft lengths. These systems have largely replaced the drag saws and other water-based processes that were prevalent a generation ago throughout the industry.

Treelength Merchandisers. Merchandising is something more than bucking specified lengths; it is a process that converts a treelength stem into log segments. The Georgia-Pacific unit at Prosperity, South Carolina, is an example.

All logs are brought in tree length to a 5-in. top and unloaded by a Raygo-Wagner wheeled loader. The logs either go into a cold deck or are fed directly onto an infeed deck to the mill. Once on the log deck, the logs proceed onto a live lineal conveyor. The logs then pass through a North American Controls scanner, which automatically activates the log hauls and stops at specified lengths for preprogrammed dollar bucking.

The resulting segments are then debarked by a 30-in. Kockums Cambio ring debarker and fed onto a surge deck. A cripple saw then cuts out all crooked and marginal material and sends it to the whole-log chipper. The logs are then segregated by diameter onto either of two log decks, which provide storage ahead of the Chip-N-Saw infeed. There is also a kickout pocket for peeler mults.

The flow at Hammermill Paper Co.'s Tuscaloosa, Alabama, mill is similar, although there are two independent lines, each fed by a Kockums 90 jib boom crane. All logs are tree length to a 6-in. top. The logs are bucked, debarked and then sent to the mill.

Simpson's Mill 5, a stud mill with some random-length capability, is illustrated in Figure 11.8. This Shelton, Washington, facility has a single infeed deck for the treelength Douglas fir or hemlock.

Figure 11.8. Treelength log processor: one conversion option.

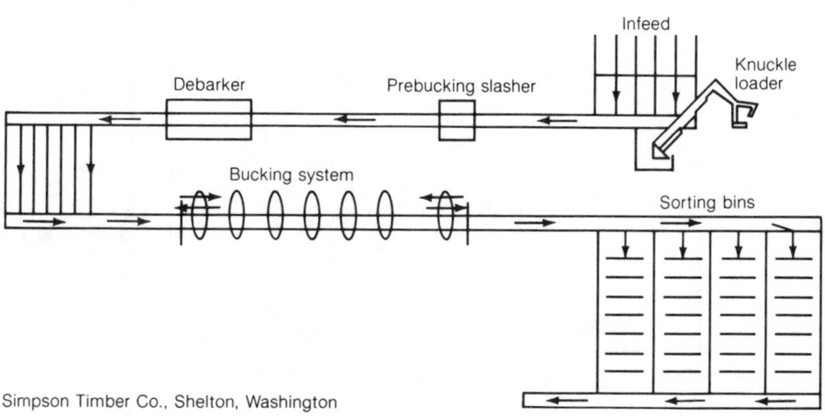

Simpson Timber Co., Shelton, Washington

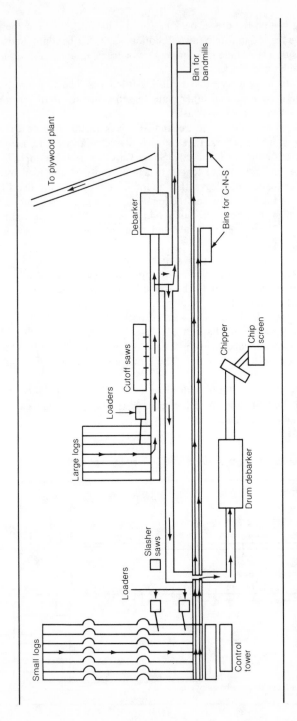

Figure 11.9. Treelength log processor: multiple conversion options.

The logs are singulated into a lineal conveyor, are carried past a 60-in. prebucking saw and then pass through a 24-in. Kockums Cambio debarker. Oversize, damaged or crooked logs are removed prior to the mill.

The logs then pass across a live surge deck and are conveyed into the mill. The incoming logs are scanned and then evaluated within Hewlett-Packard 21 MXE series computers. The program solution is determined; the software then activates a seven-saw Kockums bucking station, which in turn bucks the log into stud bolts and random lengths. The two end floating saws cut the random lengths as desired; the remaining saws are fixed at 98 in.

Downstream are four sort bins: three for stud bolts and the remaining one for the random-length 10, 12, and 14-ft lengths. Diameter sorts for the stud bolts are 4 to 6.5 in., 6.5 to 8.5, and 8.5 up to 12 in. The computers keep track of the input to each bin and then outfeed each to even the flow to the Mark II Chip-N-Saw. This sorting feature minimizes the sets and increases throughput.

Figure 11.9 is an example of a merchandiser with multiple conversion options. The logs are presorted by diameter in the woods. This merchandiser type usually requires the 10-in.-or-less butt diameter sort to be fed onto the small side. These logs are bucked at sawmill lengths, with tag ends and unmerchantable chunks going to the whole-log chipper. The large-log side feeds the plywood plant. A bin for oversize bandmill logs is included, along with a conveyor to the drum debarker and whole-log chipper for unsuitable material.

Figure 11.10 also features two lines, although each has more flexibility than the line shown in Figure 11.9. The processing line on the left side of Figure 11.10 is essentially a small-log line, although both lines can merchandise a log segment into the plywood plant, the Chip-N-Saw line or the pole yard.

This system processes about 2,880 stems per 8-hour shift with an average log length of 50 to 60 ft. A diameter sort is made previous to the mill, with the large diameters (12-in. diameter on the butt plus) going to the right-hand processing line shown in the drawing.

Weyerhaeuser, Champion, Manville and International Paper as well as others have installed similar systems in the Southeast. These systems have also been used for some years in the West Coast operations of many of these same companies. At Weyerhaeuser's Longview, Washington, and Springfield, Oregon, operations two units have been in use for many years.

West Coast processors usually require extensive dry land sorting prior to the merchandiser because of the large number of softwood species being utilized and the large number of diameter classes and quality characteristics within a species. Weyerhaeuser gleans the wood—some of which contains a high percentage of fiber unsuitable for lumber or plywood—of all usable fiber.

Merchandisers, log processors and mechanized sorters are all part of a continuing trend to less traditional log processing methods. A changing raw material resource, an evolving product market with changing customer preferences and the continuing relative scarcity of wood will ensure additional changes in the years ahead.

The successful operator will anticipate the changes—and the resulting opportunities. He will then alter methods and manufacturing procedures to take advantage of the cost and value opportunities presented.

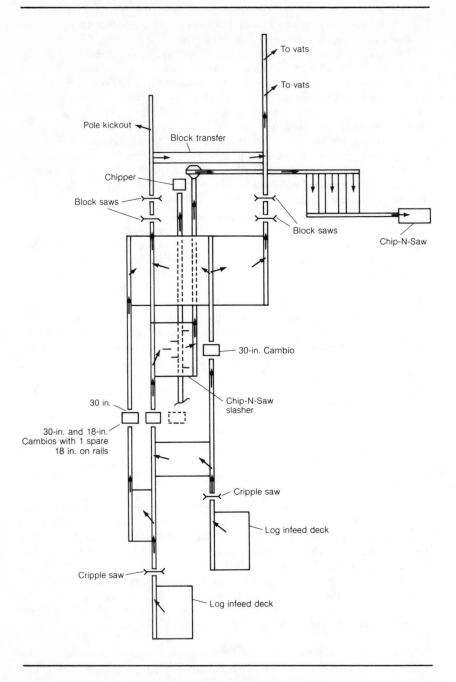

Figure 11.10. Layout of merchandiser at a southern paper company.

REFERENCES

Griffin, G. 1982. Small Log Rebirth. *Timber Processing Industry* 7(3):12.

Hampton, C. M. 1981. *Dry Land Log Handling and Sorting.* San Francisco: Miller Freeman Publications.

Heyel, C. 1979. *The VNR Concise Guide to Industrial Management.* New York: Van Nostrand Reinhold.

Holekamp, J. A. 1980. *The Efficient Harvest: Transport and Millyard Process of Smallwood Pine.* AIChE Symposium Series, vol. 76, no. 195. New York: American Institute of Chemical Engineers.

Williston E. M. 1981. *Small Log Sawmills: Profitable Product Selection, Process Design and Operation.* San Francisco: Miller Freeman Publications.

Twelve
The Softwood Manufacturing Process

Wood is unequaled as a raw material. It is uniquely suited for sawing, slicing, peeling and shaping. It can be milled into small segments and then further processed into flat panels or molded products.

Trees, especially the softwood species, are a readily renewable resource that can be furnished to nearby converting plants on a continuing basis. The lion's share of these mills converts the log into either lumber or panels. The following describes the business, the products and the manufacturing process for each.

SOFTWOOD LUMBER

The Business and the Products

The softwood lumber manufacturing business exists in a tough, cyclic business environment. It is closely tied to government policy. For example, the largest single use of softwood lumber in North America is for light framing, which is a prime component of home construction. About 85% of all housing units in the USA are of this type. When housing starts are up, business is great; when they are down, it is terrible.

"Home building and auto manufacturing will lead us out of the recession just as they have in the past," is a comment credited to one or more congressmen in each of the economic recessions since World War II.

Lumber is a global commodity. Good market conditions in one country are followed by production increases there and in other countries and often by a rise in importing. Sometimes lumber is a tool to achieve national goals.

Commented a Scandinavian lumber producer: "We can tell when the Russians need hard Western currencies. Blocks of lumber will enter the European market channels and will be sold with little apparent regard for price. We just retire from the market until it is absorbed."

Softwood lumber is an important manufactured commodity in the Northern Hemisphere, and particularly in North America. In addition, the industry continues to gain importance in the softwood-growing regions of the Southern Hemisphere, such as Australia, New Zealand and South Africa. The capacity of the manufacturing unit varies from region to region and country to country. The size of the timber, the volume density in the stands, the location of the mills, the availability of financing, and lastly, market demand are the principal variables that determine whether a mill produces a million feet a year or 200 million plus.

For example, the USA and Canada, prime lumber manufacturers, produced 44.4 billion bd ft in 1983, with over 1,050 mills producing the major portion. The top 10 producers, with 230 mills, produced about 26% of the total (Annual Lumber Review 1984, pp. 14–21). A boom housing year will see this total volume soar to 50 billion and more.

The North American producers are concentrated in the Pacific Coast area between the panhandle of Alaska and the California north coast. Major mill concentrations occur in the Willamette Valley in Oregon, in western Washington and on the lower British Columbia mainland. Also important are the USA's 13 southeastern states, where 9.6 billion bd ft was produced in 1981.

Douglas fir, white and sugar pine, hemlock, spruce, the true firs and the pine species are the more important North American softwoods. Various pine and spruce species are abundant in other Northern Hemisphere countries. The pines, in particular the radiata pine, are the predominant species found in countries of the Southern Hemisphere.

An important North American raw material, in addition to Douglas fir, is the southern yellow pine in the southeastern states. Lumber is produced from four principal tree species: loblolly, shortleaf, longleaf and slash pine. Other minor species are utilized, although each is stress graded and assigned stiffness values or visually graded and identified as a minor species.

Regional preferences for certain species are strong. These preferences are generally a function of perceived lumber strength, end user preference and/or transportation costs. Green lumber markets for desirable species such as Douglas fir are shrinking as transportation costs continue to soar. Long lengths, such as 20 to 26-ft lumber, are becoming less available as the suitable tree types become more scarce.

Factory, construction and other end-use needs are the engines that drive the manufacturing process and determine the specifications for the resulting product. Swan Alverdson, a Scandinavian, devised the first set of grading rules about 1754. Domestic grading rules appeared in use by 1833 in Maine. Grade rules followed the industry west and south during the intervening years.

North American grading rules, particularly those in the USA, are published by various regional agencies within the framework of a voluntary product standard. The latest product standard, PS 20-70, is the result of a cooperative effort of lumber manufacturers, distributors and users acting through the American Lumber Standards Committee. The Canadians also participate in the development of the American grade rules because they are a large supplier of the U.S. market.

Wood is a highly variable raw material; the rules for each grade recognize this uniqueness while identifying the just-right characteristics for a designated end use. Softwood lumber is graded as an entire piece for a combination of strength and appearance or for each of the two only.

Select, shop, common, dimension and studs are major product groupings, and there are other specialized categories. Other characteristics that must be considered include species, moisture content (green or dry measure in percent of an oven-dry condition), size and the surface finish.

North American widths commonly range from a nominal 2 to 12 in. in 2-in. increments. Lumber thickness can be categorized as follows:

Boards: Lumber less than 2 in. in nominal thickness.
Dimension: Lumber from 2 in. to, but not including, 5 in. in nominal thickness.
Timbers: Lumber 5 in. or more in nominal thickness in the least dimension.

Lumber, shown here at the dry kiln, is an important commodity.

Softwood lumber produced in North America is manufactured in 2-ft length multiples in even numbers. A random-length tally will usually include a predetermined piece or bundle count of lengths from 8 ft to 16 ft and more.

The Manufacturing Process

Lumber is produced in mills that vary widely in design and process configurations. Primary breakdown in the sawmill can be classified as one of two types: (1) grade sawing or (2) one-pass volume sawing, which is typical of the quads, scraggs, and chipping centers. Grade sawing is primarily used with bandmills and carriage rigs, which are designed to break down the larger logs that are typical of the remaining old-growth softwood species found in the Pacific Northwest and other areas of the world.

Certain processes are common to all lumber mills. These include log handling and debarking, primary breakdown, secondary breakdown, plus trimming, sorting, drying and finishing. Figure 12.1 is a schematic of the typical lumber manufacturing process.

The lumber operation has frequently been regarded as a low-technology industry, an industry that is a holdover from the past century. Innovative hardware, state-of-the-art electronics and updated management techniques have changed all that.

Today's modern lumber manufacturing process is an outgrowth of earlier innovations, innovations such as the circular saw, the gang saw and the band saw. These innovations have been updated and combined with new technology to cope with changing log types and species, along with the demands of a competitive marketplace. Bucking and debarking are the first processing steps.

Bucking and Debarking. The bucking function, increasingly computerized, seeks to maximize the value from the log by considering yield and length. Debarking, occurring either before or after bucking, removes the bark and prepares the log for sawing. Debarked logs are easier to evaluate for grade and contour, and this process also prevents bark contamination of the by-products. Large logs usually pass through a hydraulic or Rosser head barker; a Cambio or ring type is common for smaller logs. Debarked logs are then held on a conveyor pending entry onto the live deck. The primary breakdown process then occurs at the headrig.

Primary Breakdown. The typical large-log headrig is a single-band saw, either single or double cut, combined with a carriage. The carriage is usually a three or four-block platform mounted on a double track. The carriage itself is generally powered with a water or a steam shotgun, although other power systems are sometimes used.

The sawlog, evaluated as to size, contour and grade, is oriented in the proper position for cutting. The carriage blocks, or knees, are adjusted to account for taper. The sawlog is held securely as it makes a straight path through the saw. The sawyer, often aided by scanners and laser lines, evaluates the log to determine optimum grade or volume recovery.

The Potlatch mill at Warren, Arkansas, is an example large-log mill. This mill, designed for the large-diameter pine timber native to the central Arkansas area, features two large bandmills; one is a Filer & Stowell with a 51-ft band that has a

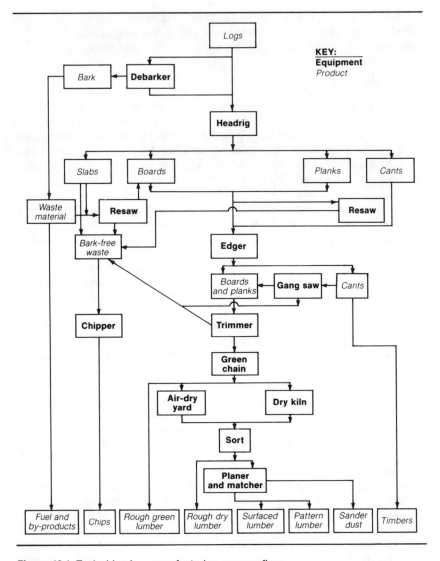

Figure 12.1. Typical lumber manufacturing process flow.

13-gauge thickness and a .195 kerf. The mill cuts for grade and wide widths. It is quite similar to its West Coast counterparts, although those mills usually have wider product mix options available to the sawyer.

The primary breakdown is where method changes are occurring at an exponential pace. Just as the pit saw gave way to the circular saw, the traditional bandmill and carriage is giving way to specialized innovative equipment such as the chipping canter and, lately, types such as the Maxi-Mill. The Fort Hill Lumber Co. at

Willamina, Oregon, is an example of this type of specialized innovation.

A computer aids the ball feed screw positioners and related hardware to position the log for end-dogging on the overhead rail. The secured log moves forward into a 6-ft Letson & Burpee quad high-strain bandmill equipped with twin CM & E chipperheads. This machine accepts logs up to 24 ft long through the high-strain bandmills, set at 13,000 to 14,000 lb. Each saw has a .125 kerf. This Maxi-Mill is a one-pass machine with a feed rate of about 3 pieces per minute. Downstream the cants and lumber flow onto a cross-transfer that in turn transports the in-process material into one of three secondary breakdown units, a Schurmann 10-in. double-arbor gang edger, a Schurmann flat edger or an Albany horizontal resaw.

While this mill produces dimension lumber, it also makes a substantial quantity of timbers. The logs themselves are a mixed lot, with both small and large diameters being cut. Short and long lengths also occur. This Maxi-Mill combined with a single-cut bandmill headrig produces about 200,000 bd ft of lumber and timbers a shift.

The Willamina log is typical of a transition forest, a forest that is changing from old growth to second growth. And much of the forest in the USA and Canada is rapidly shifting to second-growth stands where the raw material is more uniform, has minimal defects and yields predominately dimension lumber. These characteristics lend themselves to the current state-of-the-art in automated process control.

The Maxi-Mill primary breakdown unit is becoming increasingly popular.

Scanners facilitate recovery sawing by evaluating the size and contour of each log and relaying this information to the sawyer or saw control computer. A conveyor then carries the log through the bandmill, chipping canter, or some combination of circular saws, bandmills and/or chipperheads.

Chip-N-Saws are currently the most popular small-log breakdown unit. Usual infeed speed is about 180 fpm, although up to 277 fpm or more can be achieved with the installation of the frequency converter. Cants and sideboards exit downstream, with the cant proceeding into a vertical-arbor gang canter or downstream into a horizontal single-arbor or double-arbor rotary gang. Slabs with excess wane are pulled out of the flow and run through a board edger along with the sideboards developing at the Chip-N-Saw. The edger is frequently a chipping type with either adjustable sets or fixed pockets. The chipping edger eliminates the handling of unmerchantable slabs and edgings.

Small-log transport systems are many and varied. The Chip-N-Saw alligator feed system, which grasps the log, is one; side-dogging, end-dogging, sharp-chain and ducking-dog infeeds are others. An end-dogging system feeding a Maxi-Mill is an example of a system that feeds the computer-positioned log accurately into a primary breakdown unit. The years ahead will see new options and refinements of existing ones as the industry shifts to the second and third forest.

Secondary Breakdown. Equipment and technology do not vary greatly during the secondary breakdown process for small or large logs. State-of-the-art innovations have centered on feed systems, saw types (plate, profile, and cutting edges such as carbide) and process control scanning. Cants are sawn into component lumber, slabs are surfaced and edged and the resulting lumber passes to the green trimmer.

Trimming and Sorting. Downstream, each piece is trimmed for length and sorted by length and dimension. Circular saws spaced at 2-ft intervals are programmed to drop onto the untrimmed board to eliminate excessive wane, defect or odd lengths. Although sorting of lumber traditionally is a manual operation, automated systems now perform the sort function in many mills. The lumber then passes to the air-dry yard or to kilns for drying unless it will be sold green, either surfaced or rough.

Drying. Drying takes place in the open air, in a kiln or through a combination of the two. In kiln drying, the green lumber is sticker-stacked and then placed in a closed structure where heat and air flow are applied. The moisture is then driven from the wood as free and bound water is removed from the fiber. Intensive heating, circulation and venting aid the drying process. Frequently, preliminary air drying precedes kiln drying, as operators seek to reduce the high cost of energy in running the kilns.

In general, softwood lumber is dried to 19% moisture content or less. Other degrees of dryness are partially air dried (PAD), green (GRN) and 15% maximum (KD) in southern pine. With some species, such as southern yellow pine, it is almost mandatory to dry the lumber developing from the second and third forest. Improperly dried lumber has been known to "crawl right out of the lumber yard" or generally to assume unmerchantable shapes.

A typical dry kiln operation feeds the stickered unit downstream into cooling sheds and then to a sticker takedown. The sticker takedown, usually an overhead sweep or gravity drop, feeds each course onto a chain conveyor leading to the planer.

Finishing. A modern high-speed planer, either electrically or hydraulically driven, commonly runs feed speeds of about 600 fpm, although these speeds can range from 350 to 1,200 fpm. The lumber is planed to a uniform dimension, graded and sorted into packages. The graded and stamped lumber is packaged and stored to preserve its quality until shipping.

Surfaced lumber can be planed or surfaced on one side (S1S), one edge (S1E), two edges (S2E) or other combinations of sides and edges. It is usually surfaced four sides (S4S) in North America. Surfacing may be done to obtain smoothness, to achieve uniformity of size or both. This departs from the Scandinavian custom of precision-sawing during sawmilling to avoid the planing sequence.

SOFTWOOD PLYWOOD

The Business and the Products

Softwood plywood is a more recent development than lumber but has an equally colorful history. The industry had its modest beginnings in 1904, when the Paine Lumber Co. of Oshkosh, Wisconsin, constructed a plywood plant to provide panels for the Paine Lumber door. The 1905 Lewis and Clark Exposition set the stage for market development and mass production.

The Douglas fir panels exhibited at this Exposition were closely observed, and orders soon followed. Initially used as door panels, the product soon found other uses—and the market developed. Cabinets, automobile components and wallboard were a few of those uses. With the discovery of a waterproof adhesive in 1934, the uses for plywood multiplied. Total U.S. softwood plywood production climbed to 1 billion ft^2 in 1940; approximately 19.6 billion ft^2 were produced in the USA in 1983 (Pease 1984, p. 24). North American production has accounted for about 40% of total world plywood production in recent years.

Panels are currently used for roof sheathing, floor systems, wall sheathing, sidings, fixtures and other cabinet and housing applications. Present panel use averages 5,600 ft^2 (3/8 in. basis) per single-family house and 3200 ft^2 for each multi-family dwelling.

Sanded plywood is often designated Appearance grade plywood. It is used where one or both surfaces are exposed, as in shelving, cabinet doors, built-ins and furniture; it is a versatile product. The usual panel size is the standard 4x8 ft, although sizes such as 4x6, 4x7, 4x9 and 4x10 are available, with at least one U.S. operator producing 4x12 panels to order. Other widths—including 3 to 5 ft—are also available. In addition, special orders can be scarfed to size to fit customer needs.

The lion's share of the plywood produced is manufactured from softwood species, although all-hardwood panels are available. More than 70 species of wood may be used in the manufacture of structural sheathing. As with lumber, the common panel raw materials are Douglas fir and the southern pine species. CDX production makes up about 53% of all construction and industrial plywood produced in the USA.

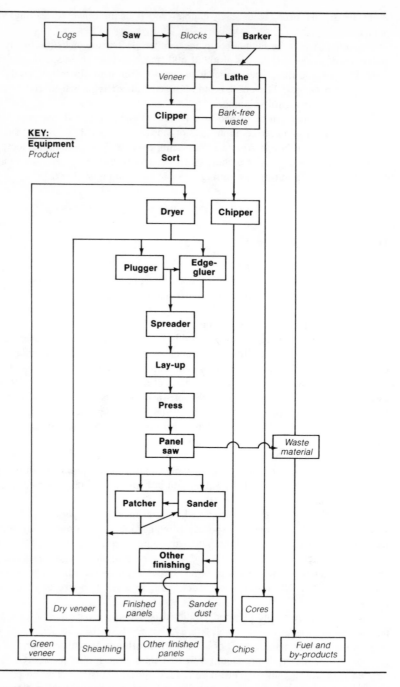

Figure 12.2. Typical veneer and plywood manufacturing process flow.

Currently, the industry's production capability has stabilized in response to the competitive inroads made by reconstituted board products plus the impact of a relatively scarce log supply. Timber costs have soared in recent years due to the apparent shortage of suitable logs.

The Manufacturing Process

The key issues for the plywood producer are yield improvement, cost reduction and value enhancement of the product mix. These issues set the pace for the manufacturing process.

A typical process, log through finished panel, is shown in Figure 12.2. The sequence of activities will vary with the mill, the product mix and the geographic area in which the plant is situated. The log is bucked, blocked and sorted for grade and length prior to peeling or prior to preconditioning before peeling.

Preconditioning. There is a wide spectrum of methods for preconditioning veneer blocks. These methods range from straight steam application to hot-water immersion, and a number of variations lie somewhere between. Each method is effective under given conditions.

The hot-water spray is most often used with spruce and white wood species, species that dominate the interior forests of British Columbia and the Pacific Northwest states. These species peel best at about 75° to 80° F.

Hot-water immersion, used as a method in the pioneer days of the industry, was reintroduced with the advent of the southern pine industry in the 1960s. The technique has since spread to preconditioning of the more buoyant species of the West Coast.

Modern preconditioning block conveying systems vary from drive-in or pass-through tunnels to hot-water immersion tanks in which the blocks are both heated and transported. The temperature and heating schedule for each method are governed by the diameter of the log, the ambient temperature and the species. The specific gravity of each species determines both the ideal peeling temperature and the rate at which a given log diameter heats. Generally the lower the specific gravity of the log, the lower the ideal peeling temperature. The conditioned blocks are then peeled on a rotary lathe.

Peeling. A large-log lathe may peel hundreds of blocks per shift; a highball small-log lathe operation may peel thousands. Each block is rotated on spindles as the lathe knife and carriage approaches at a predetermined rate. Sun Veneer, Roseburg, Oregon, exemplifies an efficient small-log operation.

Logs are sorted, transported from the dry storage area and dropped onto the storage deck leading to the ring debarker. Whole logs are debarked and then bucked into 8½-ft segments as they pass under guillotine-type twin cutoff saws. Each block then passes downstream into a sorting line.

Peelable blocks are segregated by diameter and species before steaming: logs 18 in. and larger drop into one sort; smaller blocks, into the remaining sort. The blocks are conditioned in the adjacent steam chambers, seven in number, prior to peeling.

This modern green end and others benefit from recent innovations that assist the operator to obtain more from the log while maintaining the production rate.

For example, XY chargers use scanners tied into a computer terminal to determine the largest true cylinder. The resulting information is fed to the charger control; the charger hardware then precision-spots the block onto the spindles. The powered core driver, another innovation, prevents core flexure and provides about 30% of the torque requirements as the block spins against the lathe knife. Computerized lathe controls, tray controls and clipper controls are currently used or in the process of being developed.

Lathe settings, so important to precision peeling, will too be computerized. Pressure and knife angle settings are candidates, along with knife height. The current state-of-the-art is based on the periodic use of hand-held precision instruments. The use of these instruments represents an important improvement over the hunt-and-see methods so common in the past. The state-of-the-art will further evolve into continuous monitoring and control of these functions.

Clipping. Downstream from the lathe, the direct-coupled, or single, tray to the clipper is staging a comeback. This method was used for many years as a low-cost but less productive way to transport veneer between the lathe and the clipper. The advent of small logs and the resulting short ribbons are making the usual eight-tray two-clipper operation obsolete. Increasingly the lathe that processes small logs needs only a direct-coupled system to get the job done efficiently. The veneer is subsequently clipped to width, sorted and then stacked.

Sorting. The sorting process, usually a separation of full sheets, half sheets, random widths and fishtails, also segregates some species such as Douglas fir by moisture content. This segregation, when effected accurately as a separation of the lower moisture content heartwood from the sapwood, speeds drying and improves gluing. The veneer is then dried by width, thickness and segregation.

Drying. The present veneer dryers can be categorized by air flow type, heating method and veneer transport system. Air flow is either the older longitudinal-flow or cross-flow type or the newer, more efficient impingement (jet) system. Heating methods include steam, direct-fired gas, and wood waste direct-fired burner systems. Platen drying is also commercially available.

Veneer transport methods are usually the conventional roll feed system or the newer cable slow system common in many Weyerhaeuser mills both in the Pacific Northwest and the South. A wicket system is currently being introduced in a number of installations; a lower initial capital expenditure and lower ongoing costs are its apparent advantages. Currently manufactured by Drying Systems Inc., the wicket veneer-conveying system is part of a complete dryer installation. Infeeding full sheets are grasped and rotated individually into a vertical position at the infeed end of the dryer. Each wicket in the continuous system holds a sheet securely as the veneer progresses through the drying cycle.

The dry veneer is then sorted, graded and prepared for panel assembly.

Panel Assembly. Panel assembly systems have long since been automated, although conventional glue spreaders are still popular in a number of mills. The in-line systems, such as the Georgia-Pacific/Coe and Superior types, have had widespread acceptance since being introduced in the mid-sixties. The Ashdee, or leading line, is also popular for sheathing and sanded panel construction.

Roll coating, spray coating and curtain coating are the most frequently used glue application methods, although much development work is currently being

Softwood plywood continues to be an important commodity product.

done on the foam extrusion process. Lighter spreads and greater tolerance to hot veneer are the rationale behind the development of the extrusion application.

Panel assembly costs, quality of workmanship, and relative veneer use are prime leverage points to mill profitability. Automated layup systems and more efficient adhesive application methods all play a large role in determining mill profitability.

Veneer preparation, panel patching and upgrading systems are the value-added mechanics in the mill. Veneer patching, veneer splicing, edge-gluing and the other veneer preparation machine centers upgrade the veneer to allow the production of a more desirable mix. The panel department serves the same function.

The panel saw provides the customer with a square-edged panel. The panel patching equipment, the sander, the face-texturing and edge-detail equipment plus panel scarfers and other like equipment serve one purpose: they add value to the panel mix and therefore add net incremental value to the log or purchased veneer.

STRUCTURAL PANELS: A MARRIAGE OF MARKETS AND TECHNOLOGY

Plywood will be the major element in meeting domestic structural panel needs for many years to come. It will face significant competition from other wood-based products, however, particularly those in the particleboard group (flakeboard, waferboard, fiberboard, and others).

The Business and the Products

Two plants provide examples of technology transfers between plywood and reconstituted products. The Potlatch plant at Lewiston, Idaho, has produced a composite structural panel featuring a face and back of veneer with a reconstituted-board core for some years. The Georgia-Pacific Dudley Com-Ply plant (Dudley, North Carolina) is another combination panel–producing facility.

The plant consists of two wings: one wing houses a plywood plant; the other wing houses a coreboard plant. The two processes meet at the automated layup line, where the structural panel is assembled with a veneer face and back and a coreboard inner ply.

Composite boards such as those produced at Lewiston and Dudley are developments of the 1970s. The traditional all-veneer plywood panel and the combination veneer and coreboard panel are only two of a number of panel types. Further structural panel development has led to a major change in the direction of the American Plywood Assn.

American Plywood Assn. programs have focused on veneered panels constructed within a rigid commercial standard. That commercial standard, PS 1-74, did not have the flexibility to deal effectively with the new panel types, such as the composites and the other reconstituted structural panels. The evolution to a performance standard concept has begun and will continue during the coming years.

The performance standard concept allows the producer to design the panel; subsequent test results validate performance capability. The performance results are audited and retested periodically. The design method allows the utilization of

Other structural boards, such as those manufactured at this Louisiana mill, are becoming increasingly popular. (Courtesy *Forest Industries* magazine)

both hardwood and softwood species, with the actual panel engineered to meet specified performance standards. This concept will provide:

- Improved resource utilization, both in the quantity and quality of raw material used. Fiber developed from what were thought of as weed trees can be used effectively and efficiently.
- Greater flexibility in designing a product and in putting its raw material to the highest and best use.
- More uniform performance requirements rather than narrow prescriptive specifications, which are often something more than the end user requires.

Performance-rated panels include plywood and the combination veneer-board composite panels. They also include waferboard, oriented-strand board (OSB) and other types of engineered structural boards, which are introduced periodically.

Sample Manufacturing Processes

The manufacturing process flow of a typical structural flakeboard plant is shown in Figure 12.3.

Waferboard, a structural flakeboard panel, is constructed from large, thin flakes, or wafers, generally rectangular in shape and 1 in. or more in width. Phenolic resins are blended with the wafers, and then the combined raw materials are bonded together under heat and pressure in a hot press.

A New England plant produces waferboard at Woodland, Maine. This Georgia-Pacific plant has five 165-ft-long hot ponds, which thaw and soften the fibers of the debarked blocks. Jackladders remove the 8-ft blocks, soaked 2 to 5 hours and longer at 150°F, and then the heated blocks are conveyed to a slasher saw. The resulting 4-ft lengths are fed to one of two waferizers.

One waferizer makes core wafers, which are about 1½-in. long and about twice as wide. The average thickness is about 0.025 in. The surface material is twice as long, with the thickness and width about the same as for the core material. The green wafers are then fed to a triple-pass rotary drum dryer.

The dry wafers next pass through a blender and then are conveyed to the formers. Four formers are used, two each for surface wafers and two for core. Each former lays the material in sequence on 8 by 16-ft stainless steel caul plates. The formed material and the cauls each feed to a 16-opening 8 by 16-ft Siempelkamp oil-heated press.

The hot press has the capability to obtain a maximum pressure of 750 psi at 410°F. The actual press cycle will vary from about 4 minutes for 1/4-in. board to 6½ minutes for a 7/16-in. panel.

The *oriented-strand board* (OSB) is another structural panel. This nonveneer panel is a relatively new product currently being produced in the North Central states and in the Northeast. This board consists of long and narrow wood strands, which are oriented and laid down in cross-laminated veneerlike layers. The face and back layers run parallel to the short length. The Potlatch mill at Bemidji, Minnesota, is an example.

This plant utilizes treelength aspen as furnish; each incoming stem is slashed into 106-in. lengths. Odd ends are fed to a Kockums chipper. The resulting 106-in.

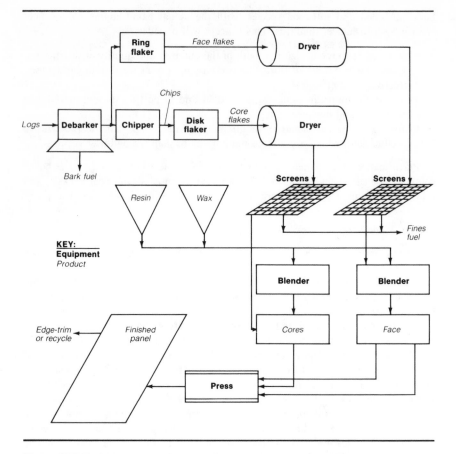

Figure 12.3. Typical structural flakeboard manufacturing process flow.

bolts are each conveyed to either of two preconditioning vats, where the fiber is thawed and initially softened. Steam injection is the heat source for these vats.

The bolts are then debarked, preconditioned further and then bucked into half lengths. These lengths then feed to either of two Hombak flakers, one for core flakes and the other for face stock. This green material then passes to triple-pass rotary dryers.

The dry fibers are fed across a scale and then move to the blending station, where they are combined with a predetermined amount of phenolic resin and other components. This prepared board furnish then passes downstream to a six-unit forming line.

Face fiber is channeled to the first and last heads in the forming line; the four inner layers are sandwiched into the center of the board. Both Lechenby and Bison formers are used as the panel is assembled on the aluminum caul plates, which feed down the continuous line into a 22-opening Washington 4 by 24-ft hot press.

The Elmendorf plant in New Hampshire is another example of a functional OSB manufacturing facility. This combination hardwood and softwood panel contains white pine, hemlock and poplar.

The incoming logs are slashed into 100-in. lengths, immersed in hot ponds for about 4 hours, cut into 50-in. lengths and then fed to Hombak flakers, which in turn develop the just-right face and core fiber.

The resulting wet fiber is transported by belt conveyor to a green flake storage bin prior to drying in a single-pass Bison dryer. The developing dry fiber is then classified into oversize, surface, core and fines. Oversize particles are further air-classified to recover all possible board furnish before being conveyed to the fuel bin.

Dry classified fiber, liquid phenolic resin and wax are mixed in the proper proportion using Bison blenders, which feature atomizing nozzles. Downstream Bison forming machines then construct the three-layer oriented mats prior to loading into a 4 by 16-ft Dieffenbacher 16-opening hot press.

OTHER RECONSTITUTED BOARD PRODUCTS

Structural reconstituted board products are new, but the board industry has a long and varied history. Hardboard, for example, was developed in 1924 by William Mason.

Sample Manufacturing Processes

The *hardboard* process subjects the incoming chip furnish to high-temperature steam in a pressure vessel. A cycle consists of approximately 600 psi of steam for about 1 minute followed by a rapid pressure increase to 1,000 psi for about 5 seconds. The pressure is suddenly released and the furnish is exhausted from the vessel. The resulting coarse mass of fibers is further reduced by an attrition mill.

The prepared furnish is then mixed with resin and other chemicals prior to being formed into a thick mat on a conveyor. The mat is then cut into panels and placed in a hot press, which in turn compresses each panel into the desired thickness. The panels are conditioned in a humidifier to restore each to a specified moisture content prior to trimming. The hardboard is either sold as a developing product or further processed as a substrate for a printed panel. Tempering each panel results in added strength and durability.

Particleboard, another reconstituted board product, is produced by combining wood particles with a resin binder. This product, commercially produced after World War II, has gained wide acceptance as a plywood substitute. Composed of randomly oriented particles, particleboard has a smooth surface that is further enhanced by sanding. It has general use as underlayment and other sheet uses.

Medium-density fiberboard (MDF), developed by Miller Hofft Inc. of Richmond, Virginia, is an unusual variation of the board manufacturing process. Hot platens and high-frequency electrical charges are used to cure the adhesive in the press. The furnish preparation process, the ingredients plus the pressing and curing cycle result in board attributes that include good strength and screw-holding capability. Smooth faces and tight edges make this an ideal panel for furniture core material.

The MDF panel is gaining wider acceptance, along with other types and variations of reconstituted boards. The production of each panel type has its key considerations and leverage points.

The hot press determines the rate of production in nearly all board plants, including the MDF process mentioned. Faster press times are being developed and are due in large measure to the development of resins that cure more quickly. Increased throughput has resulted in sharply reduced unit costs.

The fiber preparation systems are no less important. The raw material type and the fiber desired by the process are the key variables. A wide range of fiber preparation equipment has been designed and engineered. The reduction units may use knives, hammers, attrition mills or a combination of the processes.

For instance, logs are reduced into coarse chips, which in turn are fed into flakers to develop long, thin particles. Sawdust undergoes a one-step reduction in an attrition mill to develop fines for the surface layer for a smooth-face board. Planer shavings are usually processed through flakers or beater mills.

The fiber used in reconstituted boards ranges from the refined fiber utilized in hardboard and insulating board to the coarse mill flakes of the waferboard operation. The fiber preparation system varies with the board type and the properties expected in the finished product.

The manufacture of reconstituted boards, whether structural or another type, is a technology refined over 60 years of development and experience. Yet there is much to be learned in this process as well as others.

SHARED TECHNOLOGY

Increasingly there is a commonality of technology between the manufacturing processes for the various solid wood and reconstituted wood products. This is the end result of applied research combined with the evolution to small, uniform second and third-growth logs. Increased mechanization lends itself to common engineering and processing techniques. The end result is that the manufacturer must increasingly ask:

- What are my resources?
- What is the market?
- What manufacturing technique will maximize the net return on my manufacturing efforts?

The end result may be one or more of a family of products that includes lumber, plywood and the various reconstituted products. The producer will consider his company less and less as a single-product manufacturer and accept its larger role as a log and fiber converter.

REFERENCES

American Plywood Assn. 1982. *Softwood Plywood Production Statistics*. APA Management Bulletin No. FA 215. Tacoma, Wash.

Anderson, R. G. 1982. *Plywood End-Use Marketing Profiles: 1981–1983*. Tacoma, Wash.: American Plywood Assn.

Annual Lumber Review. *1984 Forest Industries.* 111(7).

Baldwin, R. F. 1981. *Plywood Manufacturing Practices.* Revised 2d ed. San Francisco: Miller Freeman Publications.

Baxter, D. G. 1977. *A Layman's Introduction to Softwood Plywood Manufacture and Practices in the Pacific Northwest.* Tacoma, Wash.: Plywood Research Foundation.

Blackman, T. 1982. New Waferboard Plant Is First to Use Softwood Material. *Forest Industries* 109(5):40.

Charles, S. 1981. Experts See Growth for MDF As Particleboard Struggles. *Plywood & Panel* 24(6):21.

Dixon, R. 1981. Elmendorf's 3-Ply OSB Receives APA Rating. *Plywood & Panel* 24(6):21.

Donnelly, D. M., and Worth, H. E. 1981. *Potential for Producing Ponderosa Pine Plywood in the Black Hills.* Resource Bulletin RM-4. Ft. Collins, Colo.: USDA Forest Service.

Finnish Paper & Timber Journal. 1980. *Wood-Based Panels in the 1980's.* Proceedings of the Symposium Organized by the Timber Economic Commission of the United Nations Economic Commission for Europe, Helsinki, Finland, May 16, 1980.

Hallock, H. 1979. Sawmilling Roots. *Electronics in the Sawmill.* Proceedings of The Electronics Workshop, Sawmill and Plywood Clinic, Portland, Oreg., March 1979. San Francisco: Miller Freeman Publications.

Haygreen, J. F., and Bower, J. L. 1982. *Forest Products and Wood Science.* Ames, Iowa: Iowa State University Press.

Maloney, T. M. 1977. *Modern Particleboard and Dry-Process Fiberboard Manufacturing.* San Francisco: Miller Freeman Publications.

Moslemi, A. A. 1974. *Particleboard. Vol. 1: Materials* and *Vol 2: Technology.* Carbondale, Ill.: Southern Illinois University Press.

Northcross, S. 1982. G. P. Continues MDF Line Run. *Plywood & Panel World* 23(8):14.

———. 1982. Potlatch Commits Minnesota to OSB. *Plywood & Panel World* 23(10):7.

Pease, D. A. 1984. Structural Panels, MDF Set Production Records. 1984 Panel Review. *Forest Industries* 111(4):24 (April).

USDA Forest Service. 1974. *Techniques For Peeling, Slicing, and Drying Veneer.* Research Paper FPL-228. Washington, D.C.: U.S. Dept. of Agriculture.

Western Wood Products Assn. 1981. *Standard Grading Rules for Western Lumber,* Portland, Oreg.

Williston, E. M. 1976. *Lumber Manufacturing: The Design and Operation of Sawmills and Planer Mills.* San Francisco: Miller Freeman Publications.

———. 1981. *Small Log Sawmills.* San Francisco: Miller Freeman Publications.

Thirteen
Mill Scheduling
and Inventory Control

Late shipments are a headache to both the producer and the customer. Fat inventories, whether in-process or finished goods, are a cash sponge. Each is usually a symptom of mill scheduling problems.

"I just don't know what I'm going to do about these late shipments. I just can't seem to break up the bottleneck in the finish end," commented a West Coast sanded producer.

The bottleneck was evident. Loads were jammed into every available space, with low grade stacked outside under tarps. Forklifts servicing individual machine centers were coping with marginal aisle clearance and tedious passageways.

"Oh, 13/16 AD? I make a lot of it," the mill production superintendent mentioned. "It is a high-volume item and it is easy to make shipments. I'll get around to the other stuff when I get ahead on volume."

The scene was little different in a dimension lumber operation in Georgia. Cited the mill manager, "I should finish the run of 2x6, 12-ft this afternoon. I'll have to get back on it on Thursday, when I work my way back into the corner of the shed. But I'll get to it sooner or later."

Mixed tiers, damaged lumber and panels, excessive machine downtime, slow-bell production and a less-than-optimum sales mix are additional costly symptoms at these two locations.

Productivity increases and low-grade decreases of 20% and more are achievable at these and other mills with improved mill scheduling. Timely shipments, smaller inventories and reduced operating capital needs are other benefits.

MILL SCHEDULING OBJECTIVES

Three profit-generating objectives are the underpinnings for a mill scheduling control system. They are:

- To obtain and maintain a relatively low cost and efficient mill operation
- To achieve and sustain a minimum inventory level

146

- To provide timely and accurate customer shipments of a consistently high quality

At first glance these objectives—particularly in view of the inventory objective—may appear to conflict. A larger finished goods inventory would appear to ensure on-time shipments. A higher in-process volume would provide longer and more efficient runs. Too often, though, a large inventory is used as a crutch to make up for poor planning.

"There's a constant trade-off, for example, between cost and quality, between short delivery times and controlled inventory" (Lubar 1981, p. 54). The task is to find the balance.

PREREQUISITES FOR SCHEDULING

The manager's role in scheduling is to harmonize the apparent conflicts between the objectives previously cited, and then orchestrate the trade-off between scheduling and inventory variables. The following are prerequisites for this task:

- The planner needs to know and understand the current management policies related to stocking and customer needs, marketing commitments and the inventory limitations of the mill or organization.
- Accurate and timely information and statistics are required. These numbers may include log yield tables, log raft or deck information, work practices and machine hours available by machine center, production rates, falldown rates and order availability.
- A can-do attitude is essential. Management personnel plus hourly employees must be committed to achieving the plan.
- Control of downtime is a must. A plan has marginal value if the mill has little control over operating rates and actual operating hours. Effective mill schedules rely on consistent machine operation.

PREPARING A PRODUCTION PLAN

There are three basic steps in preparing a production plan:

1. Determining the time period to be covered by the plan
2. Establishing the base inventory level
3. Preparing the production plan

Production-Plan Time Period

The annual plan details log availability by quantity, species and grade; identifies production capability; and outlines market opportunities. This plan is the basis for a detailed short-term production plan.

A mill production plan will be based on current orders and/or the anticipated order input. The plan will be most accurate when the time span it covers is short, such as several days or a week. An extended plan covering more than a week or

Inventory control is a prime management task.

two will be hazy unless the mill produces only a handful of commodity products from a uniform log resource.

Base Inventory Level

Product inventories exist either to support production or as a result of production. The inventory level often varies with the product type. For example, a northern California redwood mill with long drying times and numerous products will require a huge inventory compared with a mid-Willamette Valley stud mill.

The optimum inventory level is influenced by:

- The amount of effort exerted to rid the mill of small jags and falldown items
- The span of time the individual in-process loads "sit on the floor"
- How individual customer orders are worked through the mill

Clearing Small Lots. Ridding the mill of small jags and falldown items may require timely remanufacturing of the small lots. Increased floor control and more available space are the payoffs of this inventory clearing method. This strategy becomes more important when interest rates are high or in a falling product market.

Staging In-Process Loads. Fresh stock just seems to run more efficiently than in-process material that has been allowed to accumulate. Curling, bowing, staining and similar grade and handling problems develop rapidly. In-process inventory should be staged at each process step, where it can be used quickly and effectively.

The effective scheduler will set a minimum and a maximum level for each in-process item. The attainment of the minimum/maximum level will then trigger the in-process components into the next step or steps. The overall result will be a lean in-process inventory, an inventory managed much like the Japanese Just in Time inventory. The goal of the Just in Time method is to develop in house or to receive from outside vendors the just-right components at the precise time at which they are needed in the manufacturing process. Large in-process inventories are avoided.

Improved visibility between machine centers, easier tracking of in-process results and overall improved control of the mill floor are the tangible results of this method.

Limiting the Number of In-Process Items. The production of lumber or plywood usually generates a family of products. The production plan should process each developing item at a ratable flow. The mixing and fitting together of the order then occurs in the finished goods warehouse. This strategy results in increased production rates and more manageable in-process flow.

One mill, for example, develops 50 and more thicknesses and grade combinations within its extensive product mix. Each customer order generally requires seven to nine items per truckload. This manufacturer establishes and enforces a guideline specifying that no more than five to seven items are to be in the in-process flow at any given time. The demonstrated result is an even process flow with a periodic inventory buildup at the shipping dock. The overall inventory is about a third less than previous levels under a different production control system.

The Production Plan

The production plan preparation step begins with a determination of the quantity needed, by item, to fill the orders on hand or anticipated. The planner will recognize that even a crude plan is better than no plan at all and that detailed preparation will yield a more effective tool.

The resulting game plan will be the vehicle for selecting and scheduling available resources. It will track and record the process. It will also record the accuracy and timeliness with which production goals are met. Corrective action will be implemented when needed.

In addition, the plan will express the results expected at each machine center or work station in meaningful terms. Selecting the right unit of measurement is particularly important; it results in understanding at all levels of the organization.

The production plan is usually based on a block schedule method, an order scheduling plan or something in between. The schedule types function in nearly the same way in a commodity mill. The methods become increasingly differentiated when the product mix expands into a larger number of items.

The *block schedule method* groups like items together and allocates these items into preassigned time periods. The designated time periods are determined by mathematical simulation based upon expected production rates and product yields from the log or furnish.

Long runs, uniform quality levels and a stabilized rate of production are the results. The block scheduling mode is particularly well suited for mixed orders in a highly automated mill or department.

The *order scheduling plan* is defined as a method of selecting and scheduling orders in a desirable sequence through the mill. This method is as old as the industry, whether manufacturing lumber, plywood, board or other products. It was introduced when the typical mill was a relatively small series of job shops connected to the process flow by waiting inventories and carts or other means of transfer.

This scheduling type is typically used today in a mill with relatively few product variations (such as a stud mill) or in a structural sheathing mill. In addition, a low-volume producer of high-realization items (such as a cutup plant) may use this system when an exact piece count of a unique product is needed.

In each of the scheduling systems the operator must allow for expected levels of rework, falldown and scrap. In addition, the experienced mill scheduler will have one or more contingency plans as backup to the master plan. A contingency plan allows for the unexpected and is added insurance to the scheduler and customer.

The mill scheduler will also have a working knowledge of the Economic Lot Order (ELO) quantity, an idea that simply says that there is a just-right quantity at which you can obtain the best costs and the highest quality while achieving timely shipments. Individual product types will be scheduled as follows.

The Lumber Operation. There are three key scheduling points in a lumber operation. The log type or species selected along with knowledge of the product outturn is one. The cutting orders or instructions for the primary breakdown unit are another. This second point is increasingly governed by preprogrammed computerized controls. The third key point is the scheduling procedure selected to direct the green and dry lumber through the processing steps into the shipping shed.

The product mix that develops is the result of log selection and cutting practices. Processing costs and timeliness of shipments are based on manipulation of the developing lumber through the mill.

The Plywood Operation. The key to scheduling in a plywood operation is the layup department. First the layup schedule is established; then the logs and peels are programmed to meet this schedule. The orders are programmed into the layup department based on attainable veneer yields and in-mill production constraints. The veneer dryers and veneer preparation equipment, such as patchers, core formers, edge-gluers, and splicers, serve as an extension of the layup department. The finish department is then scheduled to allow for an orderly and cost-efficient flow from the layup and press area.

The Reconstituted Board Operation. The board plant lends itself to long runs of like items. Except for the cutup operation, the back end of the mill merely serves as an extension of the press line.

Each of the product manufacturing systems has a number of common features. The goal of the production scheduler is the same: to obtain the most cost effective runs, utilizing the raw material efficiently while ensuring timely shipments at the most practical inventory level. That's a tall order for any production scheduler and his co-workers. Let's further discuss practical guidelines for avoiding pitfalls.

IMPLEMENTING AND CONTROLLING THE PRODUCTION PLAN

Pitfall avoidance is almost a topic in itself, and there are a host of pitfalls to avoid. The following are some guidelines:

Build a tight accounting and reporting control system. Use timely personal inspections and spot checks as vehicles to fine-tune your information gathering system. Use the resulting numbers as discussion tools to obtain scheduling or problem-solving commitments.

Maintain a tight handle on log yields and falldown rates. Be prepared to move quickly to identify the real cause when results are less than expected.

Show-and-tell sessions are an effective communication tool. Although time-consuming, sessions of this sort should replace general conversations or intricate statistical compilations whenever and wherever possible.

Beware of expediting systems. Chances are that an expeditor is needed only if schedule achievement depends on one or two individuals. Need for an expeditor is usually an indication that communication isn't broad enough to ensure commitment by the various players.

Occasionally it may be necessary to expedite an order for a specific cause, but frequent use of an expediting system will tear down even the most effective scheduling system.

Specialize whenever possible by log type, crew, machine center or shift. A mill that makes both specialty customer-oriented products and basic commodity products will find specialization a useful tool to improve quality and reduce falldown rates, increase productivity and smooth the production flow.

Equip the dry end for maximum flexibility. For example, equip the finish department or planer shed with additional specialized process equipment; then cross-train a scant crew on a number of machines.

More equipment than employees is the rule. The mill can then use this built-in flexibility as a profit opportunity when customer demand and/or log yield indicate. The designated machine will only run when an ELO quantity is available. This strategy maximizes the much greater capital cost in mill facilities and other requisite investments in land and timber.

These guidelines were less important in the "good ole days" when logs and labor were cheap and product prices high. The scheduling function then involved little more than posting the cutting instructions or the unpriced copy of the customer order. The costs of all these things have changed over the years, and so have the mills and the operating environment. Modern scheduling methods and guidelines will help mills cope with ever more complex scheduling and customer demands.

REFERENCES

Donkersloot, R., Jr. 1981. Productivity Through Manufacturing Control. *Management Accounting*, December, pp. 25–32.

Lubar, R. 1981. Rediscovering the Factory. *Fortune*, July 13, pp. 52–64.

Fourteen
Electronic Process Control: An Overview

The forest products producer was once regarded as a country cousin of other, more progressive basic manufacturers. Acceptance of new ideas was slow; progress was measured in decades rather than business cycles. All that has changed.

The realities of harsh market downturns, skyrocketing labor costs, scarce and costly logs and the advent of computer technology have combined to accelerate the introduction and acceptance of electronic controls and other space-age technology. Passive functions, such as the automation of inventory and accounting tasks, have evolved into active functions, such as controlling machine activities and developing online data.

Computerized controls, utilizing high-tech electromechanical hardware, are bringing automation and predictability into the manufacturing process. Higher speeds and increasing levels of accuracy are occurring at an exponential pace.

ELECTRONIC PROCESS CONTROL SYSTEMS

An electronic process control system has a number of elements, as illustrated in Figure 14.1. *Sensors* gather the data; one or more *information processors* manipulate the data and then formulate a command. The command is then relayed through a *network*, which in turn activates one or more *machine control functions*. Management information is an important by-product of the system.

Sensors

A sensor is defined as "a device that responds to a physical stimulus and transmits the resulting impulse." Sensors are designed for a wide variety of applications; these applications range from relatively benign operating environments to hostile extremes of temperature, moisture, vibration and other like environmental conditions that frequently characterize the manufacturing process.

Direct sunlight and near darkness are the two illumination extremes. Sub-zero to twice the boiling point of water are the temperature extremes. Vibration, hard pounding, exposure to high and very low pH values, contact with harsh chemicals and wood extractives and even neglect are some of the other characteristics of the operating environment.

Encoders, modulated light, lasers, normal light, rays, and the human voice are some of the vehicles that transmit a stimulus to the sensor. Sensor types may be used singularly or in combination with each other.

Contact Sensors. Mechanical, or contact, sensors are used frequently; sensors of this type employ sensing arms tied to a digital encoder. The sensor arms are deflected by the passing object. The amount of deflection is measured and transmitted to the information processor.

Mechanical sensing has some inherent advantages as well as some disadvantages. Its advantages include:

- Relatively low cost
- Insensitivity (usually) to surface conditions such as moisture, color or ambient light
- Satisfactory thickness measurements when accuracy needs are not too great
- Sensors that are comparatively easier and quicker to maintain than the noncontact type

Figure 14.1. Electronic process control.

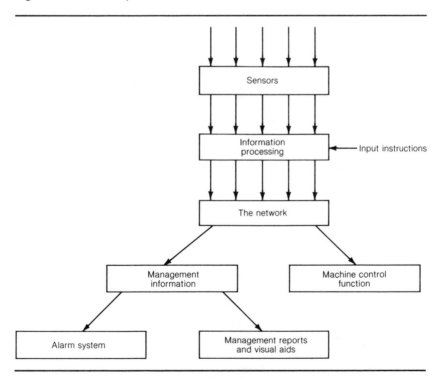

The disadvantages of mechanical sensing include:

- Lower mean time to failure than noncontact systems in similar applications
- Limited resolution for detecting surface irregularities such as roughness or offset
- Gradual degradation in the accuracy of measurement over time due to wear
- Measurement speed limited by the physical mass of the sensor

Thickness measurement at the lumber sorter, edger and trimmer are typical uses of mechanical sensing. The accuracy of the typical mechanical sensor may vary from about 0.200 in. on a sorter control sensor to about 0.020 in. on a trimmer control. The standard deviation, a measure of consistency, may vary by an order of magnitude from sensor to sensor.

Noncontact Sensors. Another sensor type is the noncontact optical sensor. Optical sensors vary greatly from vendor to vendor. They can be classified into three general groups: optical imaging systems, transmittance-type optical sensors and reflection or scatter sensors.

Optical imaging systems (Figure 14.2) typically use a Redicon tube (similar to a TV camera) or a device such as a linear array to take a picture of the passing wood. Detection is based on changes in shade or color intensity of the image. Optical imaging systems are used by Saab-Totem on computerized edgers and trimmer optimizers.

Figure 14.2. Optical imaging sensors in the sawmill.

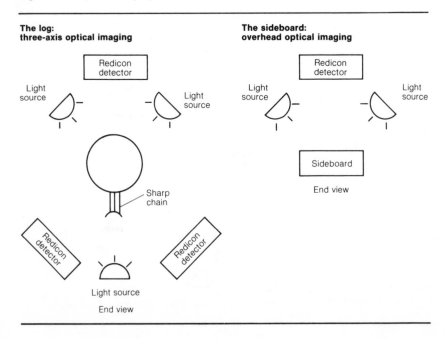

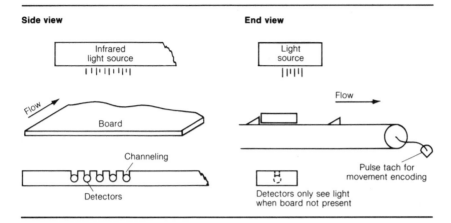

Figure 14.3. Transmittance-type sensors.

Ease of installation, low maintenance and extensive quantities of relatively precise data are favorable features. This sensor type does have some limitations: one is difficulty in measuring a three-dimensional object by scanning only one surface. Another drawback is the vast amount of data generated, which must be processed in a brief time period. These limitations have been overcome sufficiently to allow effective use.

The second optical type is the transmittance sensor. This sensor type requires a light source and a detector. Figure 14.3 illustrates a typical installation. These systems are versatile; the sensor may be used to measure dimensions or as a substitute for a switch. Opcon Corp. is the single largest supplier of this sensor type. The firm's Optimux series is widely used.

The most common light source is infrared; this is due to its insensitivity to dirt. Coe-Morvue uses a transmittance detector on its proprietary clipper control. Most primary-breakdown sawmill systems also use this detector type. The AccuRay green trimmer uses this approach for width and length measurement.

The third type of optical sensor is the reflection or scatter configuration. This type uses some form of modulated light (such as a laser) coupled with a Redicon tube or linear array (Figure 14.4). These sensors are extremely accurate and are currently available at a relatively high cost.

Several companies use this approach: Trienco with its laser thickness scanner, Coe-Morvue on its modulated light sensor for thickness sensing on an automated green trimmer, and Carl Maxey on a similar application. It is possible to achieve accuracies of $\pm$ 0.001 in. in the field with a considerable expenditure of effort and cash. An absolute measurement of thickness requires that two sensors be used simultaneously, one on the top and one on the bottom. The thickness is calculated by taking the difference between the two surface measurements. This is one reason why the approach is expensive and complicated.

Improvements in laser technology and computer power will result in price decreases for this optical sensor type. It may replace the mechanical thickness sensor

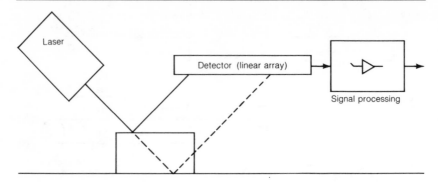

Figure 14.4. Typical reflection or scatter sensor.

by 1990. The evolution and transition in the field will be toward this and other sophisticated noncontacting sensors with accuracies in the range of 0.001 to 0.005 in.

The same sensors will be used for both process and quality control. Each sensor will be in a modular, pre-setup cartridge-type package designed to interface with high-speed digital microprocessors.

A *voice-activated sensor* is now available; this sensor will be used where human judgment is necessary, such as in subjective visual grading of lumber, veneer and panels. Raute, a Finnish company, is now marketing a veneer grading, sorting and stacking system that is activated by the voice commands of the grader/operator. This equipment features the Threshold system, a system that allows up to 250 user-defined words to command the various machine functions.

Research currently being conducted at the Forest Products Utilization Laboratory at Mississippi State University offers another exciting concept not yet commercially available to the producer. *Tomography*, currently used in medical diagnosis, is a method that mirrors and profiles the internal and external characteristics of the log. Knots and other defects are located; the resulting information can be utilized to determine the most profitable cutting pattern. The operator then has the capability to optimize the potential grade and volume yield from each log.

The development of this and other innovative sensor systems is limited only by imagination and the resources necessary to translate sound ideas into practical solutions. For example, tomography research could combine with existing technology to take most of the guesswork out of the primary conversion process.

Information Processing

Sensors feed their data into some form of information processing system, usually a digital computer or a programmable logic controller. All of the information processing systems typically used in the mill have five features in common:

- An input function
- A memory function

- A system or program for processing the input information
- A method for talking to the system
- An output function (which can range from a detailed computer printout to a full array of machine commands)

The distributed *minicomputer* is rapidly displacing the mainframe as the dominant information processing system. The newer systems feature ever smaller and more specialized processors; this trend will continue in succeeding generations of hardware. Size for size the newer systems will be more powerful and increasingly more specialized, as well as less expensive.

Thus far two dominant methods have emerged as means to program or introduce changes into the logic of computerized process control systems. The full typewriter keyboard and its variations are one; the thumbwheel method successfully used on green veneer clipper controls is another.

The *full keyboard* is standard on all computer terminals and is frequently used as the initial input device for the newer systems. It has enormous flexibility; but as an individual process control system matures, the system frequently requires fewer and fewer commands to manipulate a full range of options. The *function keypad*, a microterminal similar in appearance to a pocket calculator, frequently replaces the full keyboard. It is used by the operator to establish machine command logic, perform troubleshooting diagnostics and verify machine functions.

Another breakthrough is the *touch screen*. Several versions of this data entry system provide great flexibility by allowing the operator to enter data by simply touching a location on the cathode ray tube. There are a number of advantages to this new technology:

- Excellent operator acceptance and ease of use
- Built-in flexibility; the same hardware can be used in a wide number of locations throughout the mill

Some drawbacks include:

- Latest technology; currently very expensive
- Systems may not be rugged enough to stand up to the mill environment

The *thumbwheel*, limited as to the type and scope of information that can be input, is a further simplified variation of the microterminal. An individual thumbwheel, usually mounted in series on a control panel, represents a single machine parameter. The bounds of each machine parameter can be adjusted as the needs of the operation require. The advantage of the thumbwheel input device is ease of use; the disadvantages are its limited scope and the extensive hard wiring and numerous contacts required. If one of these contacts sticks, it will jeopardize system function.

The digital computer, relay logic panels, and solid state control systems are rapidly being replaced by the *programmable logic controller* (PLC) in a variety of end uses. The PLC performs timing, counting, data manipulation and other related functions. The complete system includes components such as limit switches, proximity switches and other sensors as input devices. A series of input/output (I/

O) racks, a processor and command implementation devices such as solenoids, motor starters and other equipment-activation devices are other system components.

A *cathode ray tube* (CRT) is used to visually inspect and confirm system parameters; troubleshooting is greatly simplified and changes can be introduced rapidly into the system. Downtime is reduced as onsite search-and-find troubleshooting inspections are replaced by electronic system scans. For example, a malfunctioning limit switch can be found and corrected in minutes rather than hours. The PLC has fixed internal wiring, but the logic functions are programmed into its memory.

The Network

The PLC and other process control tools are finding ever wider use in the mill. Individual machines are being controlled, and this control is being extended to multi-machine systems. The task is to achieve overall mill control within a coordinated data-gathering and decision-making system. The data highway is the key.

The data highway, the link between the individual process control systems, may be one of a number of types. Serial links; parallel links; bus-type links; or a combination of data, voice and video channels are some possibilities. Common to all of them is the incredible rate at which data can be transferred between process control systems and the high-capacity digital computer. This rate, which can be set at 10 megabits per second, is equivalent to about 1,400 single-spaced typewritten pages per second.

The overall mill control system will link minicomputers, programmable logic controllers and mill-floor data entry consoles to a central high-capacity digital computer, which will in turn collect, coordinate, edit and provide relevant reports or trends. The resulting information system may have analytical software capability to provide process control or management information routines such as statistical algorithms to do best fits, time and sequence correlations and other information routines.

Target size coordination, recovery tracking, saw and machine maintenance reporting, plus process alarming will be some of the benefits. Improved management communication and supervision will result as machine and process centers are tied together into an overall management control system.

There are still a number of roadblocks to the development of networks. A standard communication handshake, or protocol, needs to be developed among vendors. One vendor's equipment should have the capability to communicate with equipment from other suppliers without expensive and time-consuming development of communication devices. In addition, the network and the equipment within this network will have to be rugged enough to withstand the hostile environment typical of the log or tree conversion process.

The Electromechanical Function: Solution Implementation

Solutions can be classified as active or passive. In the passive classification, the individual must interact with the information presented to implement a solution. Usually the answer is presented on a console or another visual aid. The information may be raw statistical data, refined key indicators or more sophisticated presentations such as four-color charts, graphs or trend charts.

The Xytec microprocessor-based system is a high-technology passive system. The Model 320 TRAC (time, rate, and count) system monitors and tracks production at one or more machines. This system provides machine operators and their supervisors with accurate and timely information for up to 13 production statistics and 10 downtime categories.

AccuRay systems have wed passive systems with active process control. These systems, as part of their active machine control functions, will detect a failure condition in the system (such as a broken limit switch) and will signal the operator that something is wrong. Another system will detect suspected dollar losses (such as a too-high frequency of slashing at the trimmer in a sawmill operation). The technical limits of this AccuRay alarm system are defined only by the parameters established and the degree of sophistication desired.

The combination active/passive system described above controls the process, with the operator being little more than a spectator or supervisor. The operator's function becomes one of handling the exceptional occurrences rather than ongoing equipment manipulation.

Computerized machine control has its limitations; currently these limitations are centered at the machine. Speed and accuracy of information and decision making have little value unless machine commands can be implemented with matching speed and accuracy. Process hardware improvements, such as those available at the headrigs, veneer lathes and board lines, are receiving increased industry emphasis. The hydraulically activated linear positioner is one such development.

The linear positioner was pioneered at least two decades ago in the aerospace industry. It functions as a closed-loop system, using hydraulic cylinders to position a machine function with linear velocities of up to 40 in. per second and with accuracies to within a few thousands of an inch.

The system includes both the control portion and the feedback components; the latter identify the exact linear location of the positioner. The control portion, usually a computer-activated servo valve, precisely positions the cylinder rod at any designated increment along its length. The feedback portion usually uses a shaft encoder as a sensor and has an internal positional feedback system within the hydraulic cylinder.

There are currently two types of internal positional feedback systems: the Temposonics linear displacement transducer system and the cable reel internal feedback system. Each is capable of precise measurement.

Sawmill carriage knee positioning was the initial application of the linear positioner in the forest products industry. Veneer lathe carriage positioning plus lathe back-roll positioning are some plywood uses. Other applications are being introduced; still others are on the drawing board.

The linear positioner is just one of the innovations being developed to increase the speed and accuracy of machine commands. There will be new developments, as well as mere refinements of current electromechanical systems. A number of specialized electronic process control applications have appeared in the lumber operation, the plywood plant and the board factory. Some examples are described in the next section.

SYSTEMS IN ACTION

The Lumber Operation

"Our competitive situation cannot be materially improved by analyzing the market in search of new products or product combinations. Therefore, our efficiency to buy public timber must depend on our conversion ability to create high yields," commented industry pioneer and entrepreneur Fred Sohn (Northcross 1981, p. 11).

Sohn's stud operation at Roseburg, Oregon, utilizes computerized controls throughout the mill. The incoming log is picked up, scanned and then aligned on a one-axis mode. An end-dogging system transports the stud bolt through a Letson & Burpee twin band high-strain unit, which is combined with dual slab chippers. This mill is just an example of what is occurring industry-wide.

Computerized primary breakdown units are one of many process control applications. Log bucking and sorting, edger optimizers, trimmer optimizers, cant optimizers, kiln control and dry lumber sorting are just a few of the others.

For example, edger optimizers process the infeeding pieces through noncontact sensors, control and position each board based on a predetermined optimum solution and then feed into a flat saw edger, a chipping edger or a combination of the two. Saab/Totem, Kockums and Applied Theory Associates (ATA) supply just a few of the systems available. Yield increases of 5% and more of total mill throughput have been reported.

Computerized green trimming is also gaining wide acceptance. AccuRay, Saab Coe-Morvue, Carl Maxey and ATA units are some currently in operation. The in-

Computerized controls like this one for a green clipper introduce automation and predictability into the conversion process.

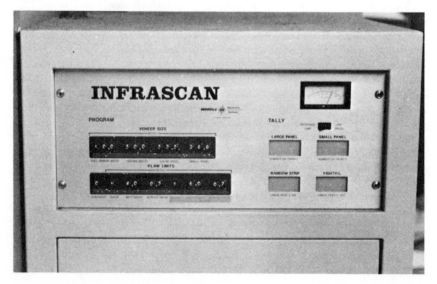

feeding lumber moves through the green trimmer sensor frame, where each piece is scanned by a contact or a noncontact sensor or a combination of the two. Command impulses initiate the saw actuators and a movable fence; the latter trims the near-side butt ends.

The operator is able to override a programmed decision, although this happens in no more than one out of 20 occurrences. Weyerhaeuser mills were the early users of the computerized green trimmer. Additional users of this equipment type and other systems are appearing at an exponential pace.

The Plywood Operation

Computerized machine control on a commercial basis was initiated with the Black Clawson green clipper control at Simpson Timber Co.'s Shelton, Washington, veneer plant in 1963. This application is so widely accepted that few manual clipping systems currently exist. Automated sorting of green veneer by moisture content segregation and width soon followed. Computerized veneer dryer control followed in rapid succession.

Recent major developments include XY lathe chargers, digital lathe carriage drives, trashgate controls, powered backup rolls and computerized carriage positioning. Tray system and green chain controls are some of the other green-end innovations. The list was extended to the dry end with the advent of panel thickness monitors, unbond and marginal glueline detectors, and the myriad applications for PLC units in the dry end as well as the green end of the mill.

The XY lathe charger is an example of what can be achieved. The equipment and its software are manufactured by Coe, Superior PMI, and others. The incoming block is scanned, realigned to obtain the largest true cylinder and then positioned precisely onto the lathe spindles.

Veneer yields of 5% and more above the norm have been documented; full sheet increases of 10% and more occur on a regular basis. The XY charger is an example of what can be accomplished with the new generation of plywood manufacturing equipment.

The Board Plant

Production of dry-formed panel products with the aid of computer-controlled forming systems is a recent innovation in the board industry. This technology, gleaned from the paper industry, is currently used in at least 20 particleboard and medium-density fiberboard (MDF) forming lines.

The La Grande, Oregon, particleboard plant of Boise Cascade sets the pace. The two forming lines are controlled by computerized instrumentation. A gamma-ray beam generated by the Americium 243 radioactive source scans the width of the mat every 14 seconds. The resulting data are processed to determine the mat weight; when the computer detects a trend away from the target weight, it adjusts the metering belt speed in the forming heads accordingly. The computer also controls the other process activities as the mat is formed prior to the hot press. The process controller effectively maintains the average weight variation within $\pm 0.5\%$ of the target.

Average board density is reduced by about 3.1% without an appreciable loss in board properties. In addition, the panel thickness variation is reduced because of a more uniform mat. Less weight and more uniform mat thickness allow press time reductions of 5% and more. Reduced wood costs, resin use and shipping weights are the prime benefits. And the customer receives an improved board.

This board example and the lumber and plywood innovations cited earlier are prime examples of ways to improve the manufacturing process. The manufacturer gains through lower costs; the buyer gains through improved quality. Additional space-age advances will further change and refine forest products manufacturing technology. In past years the manufacturing operation was more of an art than a science. The years ahead will alter all that.

REFERENCES

Baldwin, R. F. 1982. Sawmill Control—What Will the Future Bring? Paper presented at the Forest Products Research Society conference on "Computer Automation for Sawmill Profit," Norfolk, Va., October 4-6, 1982.

Northcross, S. 1981. Sun Studs Unleashes High Yield Studmill. *Timber Processing* 6(1):10-14.

Process Control in the Forest Industry. 1982. Proceedings of a symposium held November 5-7, 1980, at the Western Forestry Center, Portland, Oreg. Sponsored by Society of Wood Science & Technology, Western Forestry Center, Oregon State University.

Twedt, T. 1981. Computer Control of Forming Improves Quality, Lowers Cost. *Forest Industries* 107(11):32-33.

Section Five
MANAGEMENT AND COST CONTROL TECHNIQUES

"What are we in business for?" The immediate response: "To make lumber and plywood!" The return response: "No, it's to make money. We make money by controlling costs as we make lumber and plywood."

CHAPTER SIXTEEN

Fifteen
High-Yield Mill Management

Log and standing timber procurement is getting tougher, and each year it is getting increasingly more expensive. Add the additional requirement of securing the just-right log profile to make the desired product mix and the problem becomes a dilemma. The answer is high-yield mill management.

DEFINING THE CONCEPT

High-yield mill management is something more than a series of independent actions; it is more than occasionally measuring the product in process or selecting machine settings by "feel." It is a management system that seeks to extract the optimum net value from each log. It uses tight operating controls combined with timely state-of-the-art capital expenditures to increase profits without additional logs. The added value realized from each log then provides the leverage for coping with high log costs and marginal log quality.

The case for high-yield mill management can be summed up as follows:

Rising log costs: Stumpage prices have doubled and tripled. Stumpage sales of $500/MBM and more have not been uncommon in recent years. The resulting per-unit wood costs are frequently two-thirds and more of the sales return on the product.

A changing resource base: The log diameter is getting smaller and the piece count at the lathe or headrig higher. The mill operator must achieve the best possible yield to cope with declining volume and rising processing costs.

High-recovery equipment: Innovative high-yield equipment with relatively short payback periods is readily available. Diameter-optimizing lathe chargers, computer-controlled headrigs and other like systems can cut the log bill by 5%, 10% or more. And newer, more effective equipment is appearing each year.

Process control tools: Fact-gathering techniques and sophisticated statistical routines have been around for a long time, but the manual gathering and processing of the collected data were tedious and time consuming. Too often conditions had changed before the event had been analyzed and action taken.

Electronic devices are changing all that; sensors gather the data and the computer interprets and displays the results. Decisions based on facts can be made in minutes or even seconds.

New products: Diverse customer needs and more detailed end-user specifications are prompting new and varied products. What was once unsuitable may now be considered rustic, with premium paid by the customer.

For example, short 2x4 lengths are now manufactured into long lengths; many customers insist that end-glued short lengths make straighter studs. The price they command more than compensates for the cutting, finger-fitting and gluing. Innovative products milled or shaped from seldom-utilized species or marginal portions of the log are gaining wider customer acceptance.

Survival: High-yield mill management frequently is the only way a mill can survive the periodic downturns. Wood is the highest single cost component of the commodity solid-wood softwood product.

In the early 1970s an extensive study determined the impact that second-growth timber would have on the traditional large-log plywood mill in the Pacific Northwest. The results indicated that a 12-in. block would yield about 38% usable veneer with the then-current methods and equipment. The same log type in the same region produced about 55% and more usable veneer in a specially designed small-log mill with tight management controls.

The surviving high-yield mills of that era are higher yield mills today. The low-yield mill cited above and others like it are still waiting for the cheap log. It may be a long wait.

FACTORS BASIC TO YIELD IMPROVEMENT

Higher yields are available if the operator recognizes that:

- Increasing yield from the individual log is just part of the potential; the entire wood basket must be evaluated for conversion to additional primary products. The distinction between log types such as peelers, sawlogs, and pulpwood has become blurred. Logs down to 6 in. can be sawn into lumber; high-speed veneer lathes can profitably peel 7 to 10-in.-diameter blocks.
- The plywood and lumber manufacturing processes are unique. They result in a diminished product volume at each point at which the wood is machined, handled or stored. Realistically, all rough lumber or veneer cannot be saved.
- Loss of potential primary-product volume (waste) can be minimized through operating controls and selected capital investments.
- Operating controls and the information they produce can provide a rational basis for large gains with modest expenditures of time and resources.

Smaller diameter logs and higher piece counts characterize today's mills.

STEPS FOR DEVELOPING A YIELD IMPROVEMENT PROGRAM

Recognition of the basic factors underlies an effective yield improvement program. Activities required to develop an effective program include (1) identification of the current results at each process center, (2) comparison of those results with what is achievable, (3) definition of attainable short-term and long-term objectives, (4) implementation of management action procedures and (5) follow-up and tracking of the results achieved.

Identification of Current Results

Routine operating reports and statement comparisons will provide important clues to current results. However, a total mill survey, tailored to the individual operation, is usually required.

The survey method may range from periodic informal measurements with little actual record keeping to a more sophisticated system using electronic equipment, or it may lie somewhere in between. The performance record produced by the survey acts as a yardstick for measuring the mill's potential. The results can be a real eye opener. A comparison of small-log sawmill yields in the British Columbia interior is an example.

Six small-log mills, four with chipping headrigs and two with scraggs, were selected to represent a cross section of processing efficiency in this geographic locale. Normal mill practices were followed, although the mills bucked heavily to 16-ft lengths as part of the study requirements.

The resulting logs were sorted into four diameter classes in 2-in. increments from 4 to 10 in. The results can be summarized as follows:

- Lumber yields from small logs processed in this study averaged less than a third of the wood volume input based on cubic measurement. The smallest diameter class averaged 25% lumber; the 10-in. class was 36%.
- The between-mill variations in yields and values were substantial: The lumber recoveries ranged from 29% to 44% on a cubic basis for the same diameter class. Recovery practices were the chief cause, although there were some differences in quality.

These recovery variations are common in other regions and in other countries. The U.S. Forest Service provides other examples. After about 1,000 Sawmill Improvement Program (SIP) studies were conducted nationwide, several writers commenting on the sawing accuracy noted: "The results indicate the tremendous differences that exist between mills. For example, sawing variation ranged from ±0.025 to ±0.534 inch with a mean of ±0.165. Oversizing ranged from 0.0 to 0.288 inch with a mean of 0.077. It is obvious that almost all sawmills could benefit from closer control of sawing accuracy" (Hallock, Stern and Lewis, 1979, p. 2).

The SIP sponsored by the U.S. Forest Service is an example of a mill analysis that identifies current results. The program identifies current performance and

The surviving high-yield mills of the early 1970s—such as this central Oregon panel plant—are higher yield mills today. (Courtesy *Forest Industries* magazine)

then further identifies the potential that can be achieved with existing technology at each machine center. A like study can be conducted in either a lumber or a plywood operation.

Comparison of Results with Potential

Lumber Recovery Factor. The SIP uses statistics and ratios to compare results with what is possible. The lumber recovery factor (LRF) is a commonly used ratio. It represents nominal board feet of lumber recovered per cubic foot of log input to a sawmill. The SIP studies utilize three types of LRFs: the actual LRF, the predicted LRF and the calculated maximum LRF.

The *actual LRF* is determined by a four-step procedure as follows:

1. One hundred logs are measured inside the bark on each end, and the cubic volume is calculated.
2. These logs are sawn as a batch in a normal manner and the output is then carefully tallied.
3. The individual pieces of lumber are then measured using well-defined statistical methods.
4. The resulting log, lumber and mill statistics are classified using a software program developed for the Forest Service.

The *predicted LRF* is determined by using the software program to analyze the actual data. The computer manipulates the data and reduces the sawing variation to an attainable level, eliminates excess oversizing and reduces the planing allowance to the minimum.

The third ratio, the *calculated maximum LRF*, uses a computer sawing solution to minimize slab, edging and trim volume. The three LRFs provide the operator with both the present and potential yield opportunities, as defined by a detailed computer analysis.

Other ratios are used as performance yardsticks in lumber, as well as in plywood and board manufacture. The log rule ratio is the most common of the two described below.

Log Rule Ratio. The in-process or finished goods—defined in units of measurement such as thousand surface measure (MSM), ⅜ths basis, for plywood or veneer or thousand board measure (MBM) for lumber—are divided by the log input scale. These ratios have little value as a unit of comparison outside the individual mill because of the wide variations among log rules, the number of variations on the same basic rule, and the individual scaling methods employed.

Cubic Ratio. The cubic ratio is gaining wider acceptance. This ratio is calculated using the cubic scale as the denominator and the product unit volume as the numerator. The cubic ratio can be used as a performance yardstick on plywood, lumber and all-wood board products.

The cubic ratio; the LRF or the veneer recovery factor (VRF), which is sometimes used; and other cubic ratios are gaining increasing acceptance because of their ease of application in the various value management programs designed to measure the family of products that develop from sawtimber.

Definition of Attainable Objectives

The first step in developing attainable objectives is to define usable product, whether lumber, plywood or board. This step considers the applicable commercial standard or product requirements; it also considers what the trade will accept. A value and cost analysis is frequently required in order to determine the minimum specifications.

Second, the performance standards for each job or machine center are defined. A written performance description for the lathe operator may include a specific method for checking the charger for accuracy at specific times during the shift to prevent excessive development of fishtails or random-width veneer. In a sawmill, specific green trimming instructions may be documented and posted at the infeed to the green trimmer. The performance standard or standard operating procedure is documented and will be updated as better methods and improved equipment settings are determined.

The third step is to determine the product and by-product yield for a particular log type when utilizing specific operating practices and job performance standards. Specific log test procedures, described in Chapter 8, or other detailed methods will provide the information required when used effectively.

The detailed analysis will identify leverage points, points where a modest effort will pay off in large gains. The leverage points can be classified as follows:

Mill type and condition: Reducing sawing variation by tightening up the carriage knees is an example of a leverage point in a sawmill; tightening up the lathe carriage by replacing worn parts is its companion in veneer manufacturing. There are like examples at each point at which the log or wood segment is peeled, sawed, clipped and/or machined. The number of leverage points of this sort and the magnitude of the potential available at each will be determined by the present mill practices and the design of the mill itself.

Processing decisions: Processing decisions can be separated into capital and operating. Installing computerized controls is an example of the former; improved operating procedures (such as manual edging instructions and practices) are an example of the latter.

Product sizes: Most mills peel, saw or form too thick, too wide and sometimes too long. More precise sizes and the salvaging of smaller segments are examples of this leverage point type.

Log characteristics: Not much can be done short term to improve a tree. A whole lot can be done to improve product outturn. Selecting the right log segment for a conversion option is the key.

For example, a sound but sweepy 20-in.-dia log segment is probably better suited for the veneer plant than for the headrig. Correct bucking of the tree-length stem plus appropriate downstream conversion decisions can add lots of extra dollars; the challenge for the operator is to ascertain that each log is bucked to the correct length for the right product option.

Implementation of Management Action Procedures

The action procedures begin as the savings available are determined for each leverage point. The laundry list will provide the operator with numerous opportunities.

For example, rough green size reduction at the headrig may result in fewer planer shavings but additional lumber and pulp chips. The difference between present product and by-product value and possible value provides the gross available savings. (An example savings calculation appears in Chapter 9, Figure 9.2.) Next, the incremental processing or other cost increases or decreases are factored in to arrive at the net savings. Each leverage point is evaluated in a similar fashion and the resulting list is ranked, taking into consideration the limitations of time, money and other resources.

The first step in the ranking process is to subdivide the projects into three classifications: (1) those that are attainable through a greater emphasis on operating controls, (2) those that are considered expensed items in the budget and (3) those that require capital expenditures.

The first classification emphasizes setting performance standards and then achieving these standards through training and follow-up. The second classification is also straightforward; resources are obtained within the limits of available maintenance or supply funds considering company and Internal Revenue Service spending definitions. An improved thin-kerf saw design or a saw guide system are examples of the second category. These relatively small improvements, designed to decrease saw kerf and sawing variation, are examples of small projects that can pay off in a big way.

Capital projects are the third classification. Projects of this sort can include replacement of an edger; they may include a more accurate rotary green clipper for the veneer line. The objective is to obtain the greatest return on investment for the money spent. (See Chapter 9 for information on analyzing the rate of return.)

The projects are classified and ranked in the order of importance. The operator or manager then determines a programmed course of action by designating who does what by when at what cost with what expected results. The plan is then communicated and implemented.

Follow-Up and Tracking of Results

The objective of a yield tracking system is to collect accurate data within the time span desired. Yield control information can be developed for one or all of the following three time spans (Figure 15.1): the period immediately preceding the process (standard yield accounting), the period during the process (yield accounting) and the period after the process (yield finding).

Standard yield accounting prepares yield standards based on specific operating criteria for a particular log type. Product yield is calculated prior to processing the log into products.

Yield accounting, conducted during the process, measures the ongoing conversion activity at each machine center as part of the ongoing production process. Corrective action can then be taken on a timely basis.

Yield finding is an overview of the results achieved. Yield is calculated using

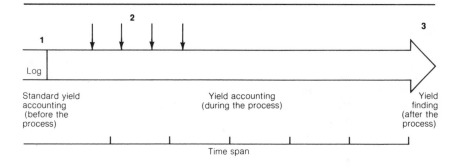

1. *Standard yield accounting* determines the standard volume of product that should develop, based on a specific set of operating criteria.

2. *Yield accounting* is similar to direct costing in accounting. Yield is measured completely at a process center. This method measures the actual product developing.

3. *Yield finding* determines the total product developed during a time period.

Figure 15.1. Yield tracking system. (Source: Modified from Baldwin 1981, p. 55)

one or more of the ratios outlined earlier. The resulting recovery ratio represents a broad average that reveals little about performance at the individual machine centers. But it does provide an overall scorecard that reflects the per-unit wood cost which will be documented on the operating statement. Logs processed are compared with the weighted average of current log test results to determine how well the mill is performing overall.

The yield controls described take the guesswork out of the process; each is a part of the framework of systematic tracking and follow-up procedures that provides the manager with the facts to make effective decisions.

OTHER ADVANTAGES OF A YIELD IMPROVEMENT PROGRAM

A yield improvement program can help a mill operator or supervisor take the guesswork out of process control. With yield accounting, problems are identified as they occur. Time otherwise spent searching for yield-reducing factors can be better spent in improving those factors when they are quickly identified. In addition, monitoring of the ongoing process can assist mutual understanding at all levels of the organization. An improvement in methods or equipment is more likely to result in an improved operating statement. And an improved operating statement can provide the incentive for further yield improvements based on available resources of men, machinery and capital.

The 1980s will see an intensified quest for more efficient log utilization. Increasing log costs and a greater demand for forest products emphasize this trend. More efficient log utilization will be the cushion that can effectively absorb rising wood and other production costs.

REFERENCES

Baldwin, R. F. 1981. *Plywood Manufacturing Practices*. Rev. 2d ed., Ch. 5. San Francisco: Miller Freeman Publications.

Clapp, V. W. 1980. Improving Lumber Recovery in the Pacific Northwest—Some SIP Experiences. Speech given at Ed Williston's Lumber Manufacturing Workshop, Seattle, Wash., Portland, Oreg.: USDA Forest Service, PNW Region 6.

Clephane, T. P. 1981. The Favorable Outlook for Southern Timber. Speech delivered to the 40th Southern Forestry Convention of the Forest Farmers Assn., Williamsburg, Va. Morgan Stanley & Co. Inc., New York.

Dobie, J. 1978. *Small-Log Yields in the B.C. Interior*. Information Report VP-X-167. Vancouver, B.C.: Western Forest Products Laboratory.

Dobie, J., and Wright, D. M., 1970. The Lumber Recovery Factor Revisited. Reprint from *The British Columbia Lumberman*, Vol. 54, no. 5 (May).

Fahey, T. D. 1974. *Veneer Recovery from Second-Growth Douglas-Fir*. Research Paper PNW-173. Portland, Oreg.: Pacific Northwest Forest and Range Experiment Station, USDA Forest Service.

Fahey, T. D., and Martin, D. C. 1974. *Lumber Recovery from Second-Growth Douglas-Fir*. Abstract of Research Paper PNW-177. Portland, Oreg.: Pacific Northwest Forest and Range Experiment Station, USDA Forest Service.

Hallock, H.; Stern, A. R.; and Lewis, D. W. 1979. *Improved Sawing Accuracy Does Help*. Research Paper FPL-320. Madison, Wis.: Forest Products Laboratory, USDA Forest Service.

Lane, P. H.; Henley, J. W.; Woodfin, R. O., Jr.; and Plank, M. E. 1973. *Lumber Recovery from Old-Growth Coast Douglas-Fir*. Research Paper PNW-154. Portland, Oreg.: Pacific Northwest Forest and Range Experiment Station, USDA Forest Service.

Pnevmaticos, S. M.; Corneau, Y.; and Kerr, R. C. 1981. *Yield and Productivity in Processing Tree-Length Softwoods*. Technical Report 507E. Ottawa, Ont.: Eastern Forest Products Laboratory, Forintek Canada Corp.

USDA Forest Service. 1973. *Increasing Your Lumber Recovery: Sawmill Improvement Program*. Washington, D.C.

Sixteen
Cost Control and Productivity

A producer can't change his log costs much over the short term, and he can't control market prices. But he can control those costs that are directly related to the manufacturing process. This chapter focuses on the process of controlling the controllable. The goal is to introduce the reader to practical tools that will shave costs and increase profits.

COST CONTROL PRINCIPLES

Cost control principles are many and varied. Each manager will identify those that are the most useful, and then he will add others that match his management style. The following are examples of some common precepts for controlling costs.

"It's Gotta Fall to the Bottom Line" Principle

"If you spend it, it can't fall to the bottom line," a lumber manufacturer mentioned as he spread a computer printout containing last month's results before the listener. He further explained as he thumbed through the printout, "Those sales dollars . . . if you let them stop here, here, or here . . . they just won't make it to the bottom line. We kept the mill running but we didn't let many stop in January."

January was a record profit month for this operator, a record month when most of his better capitalized counterparts had curtailed or shut down operations. This operator uses the It's Gotta Fall principle to profit and to control costs. "If you spend it, it just can't fall to the bottom line!"

Numbers Principle

"If it doesn't appear in the statement, it never happened" is the crux of the Numbers Principle. All interim reports—whether shift, operating day or weekly—should be bench-mark reports. These reports are an accurate statement of results achieved and a solid indicator of what the month-end statement will show if a specific trend continues. Higher production rates will be translated into lower unit or total costs; a yield improvement at one machine will be reflected on the statement.

The monthly statement becomes a narrative of what was allowed to happen rather than a historical statement of what just happened. The Numbers Principle has a close cousin, the People Performance Principle.

People Performance Principle

While the Numbers Principle focuses on record keeping and maintaining the score-card, the People Performance Principle zeros in on individual and team performance through the use of control reports.

A well-prepared control report will pinpoint trends and variances, both favorable and unfavorable. The time the manager spends on the floor and in the field is used more effectively in teaching and in follow-up rather than in searching and wondering. Unfavorable trends are corrected before each can affect the monthly statement; favorable trends are tracked closely, with control bench-marks established at periodic intervals of progress.

Pocketbook Principle

Controlling the authority to commit or spend funds can be called the Pocketbook Principle. The cost-conscious operator will minimize the number of personnel who have the authority to commit funds. The spending authority of those who can commit funds will be no more than absolutely needed. Minimizing the hands on the pocketbook is an effective way to control spending. An extension of this principle—the Time and Tracking Principle—simplifies and controls spending.

Time and Tracking Principle

As a rule of thumb, the greater the number of expenditures, the greater the management time required to manage the spending. The preparation of requisitions and purchase orders and the individual tracking and receiving activities are time consuming and costly for the purchasing personnel. The job can be a lot easier for the manager and his people if the number of individual purchases is sharply curtailed. Purchasing practices are modified to achieve this goal. Economic lot quantities (see Chapter 13) are ordered rather than small lots; multiple purchases are grouped onto a single purchase order. Mill use and stocking practices are closely monitored to prevent waste or unnecessary use.

The Time and Tracking Principle has as its theme, "If you don't spend it, you don't have to track it." The result is a financial report and supporting-document package that is skinnier in both outlays and paper.

Indicator Principle

Some costs are just indicators of sloppy management; these costs include demurrage, product claims, and fines for environmental or safety infractions. Each has little to do with the actual cost of manufacturing. Eliminating these and other like costs goes beyond cost control; it provides a signal to others that the cost control program is a serious, ongoing activity.

Some years ago this question was asked in a supervisor's meeting: "What are we in business for?" The immediate response was, "To make lumber and plywood!" The return response: "No, it's to make money. We make money by controlling costs as we make lumber and plywood." This is the underlying theme of all the cost control principles, principles that provide a demanding yet motivating climate for increased profits.

THE MECHANICS OF COST CONTROL

The manufacturing manager is not expected to be a cost accountant, although sometimes that would be a help. He is expected to understand the mechanics of cost control, which provide the framework for applying the principles previously mentioned.

Definitions

The effectiveness of individual cost control techniques can be judged using two criteria: the direction that the methods provide and the intensity of those directions. The following are time-tested definitions that define both criteria.

Cost Control. Cost control is the activity a manager performs to obtain a handle on the money being spent and to determine what should be spent. It also describes the activities used to make the "should be" happen.

The manager tracks progress or resolves unfavorable trends; his goal is to search out and implement improvements to his plan, or budget.

Budget. A budget identifies unit costs as well as total dollar costs, with a breakdown that is sufficient for the control task. Historical costs, expected costs (which factor in anticipated cost and productivity increases), or standard costs are typically used.

Standard costs are usually based on engineered performance standards and zero-based costing. This approach provides total accountability when the numbers are prepared using agreed-upon cost and performance norms.

Cost Centers. A cost center is the smallest segment of an activity or area of responsibility for which costs are accumulated. Cost planning and budgeting, cost accumulation and cost control activities all focus on individual cost centers. A cost center may be defined as a single machine, a grouping of machines or even a small department.

Cost Types. Costs are relative. The traditional designations of *fixed costs* (those that do not change in total as the rate of output varies) and *variable costs* (those that change in relation to changes in the rate of output) are valid as long as the manager has a feel for the relevant range for which the traditional definitions are accurate.

Understanding cost behavior is particularly important. The forest products manager is constantly faced with a cyclic demand and with rapid and volatile price, cost and operating schedule changes.

Table 16.1. Plywood Plant Cost Analysis: U.S. West Coast Average, 1980

Cost	Per M3/8ths	Total ($000)	% of Total
Direct labor			
Green end	$13.16	$1,110	6.8
Drying	13.04	1,099	6.7
Edge-gluing	1.45	122	.8
Veneer patching	5.15	434	3.0
Layup and pressing	11.77	992	6.1
Panel trimming	.90	76	.4
Panel sanding	1.39	117	.7
Panel patching	2.80	236	1.4
Warehouse and shipping	.97	82	.5
Other	.73	62	.3
Subtotal	51.36	4,330	26.7
Wood	104.23		
Core sales (net)	(4.57)		
Chip sales (net)	(17.13)		
Miscellaneous costs[1]	5.60		
Subtotal	88.13	7,429	45.8
Glue	8.14	686	4.2
Other direct material	3.06	258	1.6
Subtotal	11.20	944	5.8
Total direct costs	**150.69**	**12,703**	**78.3**
Plant overhead			
Supervision	3.90		
Indirect labor	1.28		
Maintenance labor	6.15		
Power and fuel	7.49		
Operating supplies	4.96		
Maintenance supplies	4.51		
Subtotal	28.29	2,385	14.8
Plant fixed costs			
Taxes	1.01		
Insurance	.58		
Depreciation	4.20		
Subtotal	5.79	488	3.0
General and administration	7.52	634	3.9
Total cost	**$192.29**	**$16,210**	**100.0**

Average annual 3/8ths production (M): 84,298

Source: American Plywood Assn. 1981
[1]Veneer purchases, sales and inventory adjustments

Example Cost Analysis

Table 16.1 is an example of a cost analysis based on the cost structure for the average U.S. West Coast softwood plywood plant in 1980. The per-thousand and total costs are identified for each line item; direct labor is identified for each cost center. The other cost ingredients are handled as either direct or indirect cost items. A further refinement of this cost analysis would involve specifically identifying many of the so-called indirect costs or specific direct costs for a given machine or cost center. Maintenance labor, maintenance supplies and operating supplies along with fuel and power are examples of these costs.

Note that direct costs represent 78.3% of the total costs. Four line items—the direct labor costs for the green end, drying and layup, plus wood costs—represent about two-thirds of the total direct costs. This is generally true for most plywood plants. The pattern is frequently similar in softwood converting units such as lumber and board.

Exception Reporting Tools

Analysis of Variances. Figure 16.1 illustrates how a manager can focus in on the individual line items to determine what opportunities are available when comparing planned to actual costs. The calculations are based on the costs identified in Table 16.1.

In Example 1 the direct labor for the green end has been identified for a month period. This illustration shows the per-thousand and total costs for both the plan and the actual results. The right-hand column identifies the variance from plan. An analysis will define the reason for the variance.

The total variance is $11,700. The volume exceeded plan, resulting in a $1,100 variance. But the real culprits are overstaffing, long hours and perhaps the improper rate for several or more jobs—resulting in a $10,600 unfavorable variance. Based on this analysis the manager would audit the operating hours, the crew count and the pay rate of each. He would systematically trace the variance back to the cause and prevent future occurrences.

Example 2 is a comparison to plan for the actual wood costs. It highlights a total variance of $28,200 for the same time period. To analyze the variance the manager recognizes the prime factors: volume, yield, and log grade or type. Greater-than-plan volume developed a $6,900 variance. An unfavorable recovery ratio resulted in an $8,300 variance. The manager also notes that the log cost variance of $13,000 represents the lion's share of the variance. He then may discover that more No. 3 peelers were consumed than planned. He then determines whether the richer mix can be justified on the basis of a more profitable product mix, whether lower cost logs were unavailable or whether the raft or ramp selection was incorrectly made.

The variances illustrated focus on exception reporting. The approach assists the manager to focus on the vital few and provides the framework to formulate detailed questions that will lead to the facts. The facts form the basis for decision making and follow-up.

The control examples shown are based on a monthly accounting cycle. This cycle is typically too long for effective floor control; many mills use a weekly cycle with good results. This requires the weekly gathering of the costs plus a weekly physical inventory of logs, supplies and in-process and finished goods to ascertain accurately the per-thousand and total costs for the time period.

Figure 16.1. The variances—exception reporting of performance: two plywood plant examples.

Example 1: Green End Direct Labor

Cost category	Plan		Actual		Variance	
	$/M3/8ths	Total ($000)	$/M3/8ths	Total ($000)	$/M3/8ths	Total ($000)
Direct labor Green end	13.16	92.4	14.65	104.1	1.49	11.7
Production (M3/8ths)		7,025		7,109		84

Variance due to:

Volume (Difference in quantity × Planned cost)
84 M3/8ths × $13.16   = $ 1.1 M

Cost (Difference in cost × Actual quantity)
7,109 M3/8ths × 1.49   =   10.6 M

Total   **$11.7 M**

Example 2: Wood Cost

Cost category	Plan		Actual		Variance	
	$/M3/8ths	Total ($000)	$/M3/8ths	Total ($000)	$/M3/8ths	Total ($000)
Wood cost in veneer	82.00	576.0	85.00	604.2	3.00	28.2
Production (M3/8ths)		7,025		7,109		84
Recovery ratio		2.83		2.79		0.04

Variance due to:

Volume (Planned cost × Difference in quantity)
$82.00/M3/8ths × 84 M3/8ths   = $ 6.9 M

Yield (Planned wood cost × Difference in veneer yield × Actual production)

$82.00/M3/8ths $\times \dfrac{2.83 \text{ (Plan RR)}}{2.79 \text{ (Actual RR)}}$ = $83.17

($83.17 − $82.00) × 7,109 M3/8ths   =   8.3 M

Log Cost (Actual wood cost less Planned wood cost revised by Actual yield × Actual production)
($85.00 − $83.17) × 7,109 M3/8ths   =   13.0 M

Total Variance   **$28.2 M**

Productivity Profile. Another exception reporting tool is the productivity profile. Each profile can be tailored to highlight the results and trends of an operation so that they can be compared with others. This profile, an index or ratio, typically focuses on the big-ticket items such as logs and labor. It can be used for equipment, material or facilities use as well as other like applications. The index is usually a ratio of output to input. The log ratio described in Chapter 15 is an example of one such ratio or index.

Figure 16.2 is an example of log and labor ratios used in a plywood plant, but they are applicable in other manufacturing operations. Example 1, a log ratio, is a determination of the product produced per unit of log input. This plywood example identifies 2.83 M ⅜ths of plywood produced per MBM of log input to the lathe based on a Scribner scale.

A similar format can be used in a lumber operation, a board plant or even a paper mill simply by dividing the product produced by the raw material input. Cubic volume is increasingly used as a denominator in calculating the ratio for all solid-wood converting operations.

Example 2 in Figure 16.2 is similar in format. It calculates labor efficiency, either direct or total, by using the paid man-hours as the numerator or the denominator when calculating the unit output. The two labor profiles shown are merely reciprocals of each other. The one selected for use is usually based on the personal preference of the manager.

Ratios can be calculated for all sorts of situations, but their usefulness will depend to a great degree on how they are applied. A ratio or profile can be applied to log or labor utilization at specific points in the process. It also can be related to use of resins, waxes, saws and knives, and other materials. It can be used for between-mill comparisons, although the user must be cautious in applying a direct comparison until both the similarities and disparities between the plants being compared are identified.

Figure 16.2. The ratios—a profile of efficiency: two plywood plant examples. (Source: American Plywood Assn. 1981)

Example 1: Wood utilization efficiency

Net M3/8ths/log rule

$$\frac{84,298 \text{ M3/8ths}}{29,787 \text{ MBM}} = 2.83 \text{ recovery ratio}$$

Example 2: Direct labor efficiency

Units/man-hour (M3/8ths/paid direct man-hour)

$$\frac{84,298 \text{ M3/8ths}}{344,073 \text{ paid m/hr}} = 245$$

Man-hours/M units (paid direct man-hour/M3/8ths)

$$\frac{344,073 \text{ paid m/hr}}{84,298 \text{ M3/8ths}} \times 1,000 = 4.08$$

Note: Based on actual number plus an estimate of direct labor man-hours.

Labor Cost Reduction Steps

As illustrated in Table 16.1 direct and indirect labor represents nearly a third of the total cost of plywood produced on the U.S. West Coast. The total is significant for each operator, whether it is more or less than the illustrated cost. In addition, for every labor dollar spent there are hidden costs, some of which are significant. Labor and labor-associated costs are usually a prime cost-reduction opportunity.

Step 1. The first cost reduction step involves an evaluation of present crewing practices within the operation. There are numerous causes for excess labor costs;

Figure 16.3. Daily crew report.

Date: MARCH 14, 1984

Department	Shift 1 Budget	Shift 1 Actual	Shift 2 Budget	Shift 2 Actual	Total Budget	Total Actual
Sawmill						
Yard	3	3	2	2	5	5
Debarker	1	1	1	1	2	2
Cutoff saw	2	2	2	2	4	4
Headrig	2	2	2	2	4	4
Edger	1	1	1	1	2	2
Trimmer	2	2	2	2	4	4
Resaw	1	1	1	1	2	2
Gang saw	1	1	1	1	2	2
Separators	2	2	2	2	4	4
Reman edger	1	1	1	1	2	2
Conveyor chaser	1	2	1	1	2	3
Hourly man	2	2	1	1	3	3
Sorter	3	3	3	3	6	6
Forklift	1	1	1	1	2	2
Utility	3	3	2	2	5	5
Total Actual Crew:	26	27	23	23	49	50
Absentees:		2				

most mills contain at least several. Evaluating crewing practices prompts a number of questions.

Are the work stations designed with labor-saving devices? Will an improved flow of in-process goods allow two or more jobs to be combined into a handyman job? Is the crew inflated to meet apparent need, such as excessive absenteeism? What effect is absenteeism having on crew and overtime replacement costs?

Step 2. The second step is to zero-base the man-hours and crew count to the expected volume and mix of products. The man-hours needed to meet mill volume should take into consideration the expected training requirement, the forecasted absentee rate, needed maintenance, and necessary cleanup and housekeeping.

Step 3. Figures 16.3 and 16.4 illustrate in sequence a format that can assist the operator in gaining control of the man-hours after the basic crew count and man-

Figure 16.4. Man-hour report.

Week: MARCH 1, 1984

Department	Shift	Mon	Tues	Wed	Thurs	Fri	Sat	Total	Budget Total oper. day
				Actual man-hours					
Sawmill	1	238	—	—	—	—	—	—	229
	2	192	—	—	—	—	—	—	193
Dry kilns	1	24	—	—	—	—	—	—	24
	2	8	—	—	—	—	—	—	8
Planer	1	209	—	—	—	—	—	—	209
	2	209	—	—	—	—	—	—	209
Shipping	1	16	—	—	—	—	—	—	16
	2	8	—	—	—	—	—	—	8
Maintenance and power	1	24	—	—	—	—	—	—	24
	2	24	—	—	—	—	—	—	24
	3	24	—	—	—	—	—	—	24
General	1	16	—	—	—	—	—	—	16
	2	8	—	—	—	—	—	—	8
Total man-hours:		1000							992
Daily production (MBM):		225							230
Man-hours/MBM:		4.44							4.31

hours have been established. The format shown recognizes the key role of the line supervisor in effecting improvement and is designed to correlate closely with the payroll sheet. A daily man-hour report and a determination of the productivity profile (man-hours per MBM) completes the format.

Figure 16.3 illustrates a basic crew report by shift. The report lists only the sawmill crew; a complete report would include kiln, planer, and shipping as well as maintenance and by-product personnel. At shift end, the supervisor fills in the report and compares the figures to the standard budgeted crew. A deviation shows up clearly.

Figure 16.3, a man-hour report, or a like report is then prepared. Often the two reports illustrated in Figures 16.3 and 16.4 are combined into one consolidated report. Figure 16.4 is filled out by the supervisor or a designated clerk. The recorded man-hours should match those on the timesheet.

The daily crew report identifies and allows comparison of the allotted or budgeted hours with the actual. The daily man-hour compilation and the daily production calculation provide the basis for figuring the direct labor cost per thousand. For example, this calculation may indicate a 4.44 man-hour figure. The average fully loaded hourly rate for the mill may be $11 per hour of direct labor. This labor cost multiplied by the unit productivity provides a tool to measure mill costs in relation to the sales return.

The illustrated format is simple, and it can, when properly used, save thousands of dollars by preventing unpleasant surprises at month's end. It also provides a short-term yardstick to measure improvement. It assists in supplying the base information that can, in turn, result in more accurate budgeting. It results in more efficient use of the mill's most valued resource: its people. Such a gain is a bonus to effective cost control. Ultimately, the success of a cost control program depends on timely and effective people response.

REFERENCES

American Plywood Assn. 1981. *Annual Cost Report for the West Coast Plywood Industry, Year 1980*. Tacoma, Wash.: Cost Accounting and Statistical Services Dept., APA.

Babson, S. M., Jr. 1981. Profiling Your Productivity. *Management Accounting*, December: 13–17.

Baldwin, R. F. 1972. Three Ways to Control Labor Costs. *Wood & Wood Products* 77(5):28 (May).

Horgren, C. 1967. *Cost Accounting, A Managerial Emphasis*. Englewood Cliffs, N.J.: Prentice-Hall.

Murray, D. R. 1981. How Management Accountants Can Make a Manufacturing Control System More Effective. *Management Accounting*, July: 25.

Seventeen
The Role of Quality Control

Quality assurance, grade checking, product certification, quality management, and process control are all terms used to describe quality control or its related activities. Quality control is defined as "a systematic way of guaranteeing that organized activities happen the way they are planned" (Crosby 1979, p. 22).

THE BENEFITS

A formal quality control program has a number of benefits. These can be identified as follows.

The net value from the log is increased. Doing the just-right activities at the just-right time during the manufacturing process will result in more primary product from the log. Waste is minimized, sizes are reduced to the minimum needed and the finished product is just right for its intended use.

A wider and more valuable product mix is attainable. A basic product, whether lumber, plywood, board or some other forest product, can often be satisfactorily upgraded into a specialty product that requires more detailed and generally tighter specifications.

More diverse and often marginal logs and furnish can be satisfactorily milled. Often the difference between using or not using a log type or species is the difference between controlling or not controlling the manufacturing variables inherent in the primary milling, drying and other manufacturing steps for the particular product. Quality control can permit substitution of marginal log types and species for the traditional sawlogs and peelers. Old growth is replaced by the second and third forest; hemlock and aspen stand in for more desirable species.

The management task becomes clearer and more results oriented. "We decided we were in the business of causing and measuring conformance to the requirements. Therefore, quality means conformance. Nonquality is nonconformance.

"Suddenly the whole thing becomes clear. Instead of thinking of quality in terms of goodness or desirability we are looking at it as a means of meeting requirements" (Crosby 1979, p. 45).

Product liability risk is reduced. The product liability risk is twofold. First is litigation. Currently everybody sues anybody, and nobody is left out as the disgruntled customer seeks the deep pocket. Even an out-of-court settlement can be costly.

Secondly, products are being engineered and designed closer to the minimum parameters for the specified end use. The customer is no longer willing to pay the cost of extra wood, labor and materials that have heretofore been part of the finished product. Updated commercial standards and other manufacturing specifications, such as those for Performance-Rated Panels (PRP), are designed and tested to ever narrower specifications.

Records are prepared as a result of an effective quality control program and are the framework for ongoing tracking and control. These records provide an invaluable tool to counter customer complaints that arise from misuse, misuse that can no longer be ignored as we move to more engineered products. "Misuse becomes potentially catastrophic, such as found with truss failures" (Maloney 1983, p. 10).

PROGRAM OBJECTIVES

Byrne Manson of Simpson Timber Co. stresses management control in citing specific goals for a quality control program. "The purpose is to hold all elements of a business within their most economic limits and to quickly highlight those areas needing management attention. Conversely, it should identify those elements which are at a satisfactory level so that management time will not be wasted on them" (Fransworth 1980, p. 1).

Another writer describes the objective as not one of "assembling a huge staff at headquarters for the purpose of strangling every potential problem in its crib" (Crosby 1979, p. 7). An effective program will be a grassroots program, a program designed to function as a management and process control tool for the operator and the supervisor. Clearly spelled out definitions of program elements are essential to successful quality control.

PROGRAM ELEMENTS

What makes up a basic quality control program? Essentially, it is a systematic method of determining quality requirements and procedures, combined with a data collection system that includes calculations, report processing, data analysis and timely follow-up as elements that can lead to process modification and correction. A description of each element follows.

Quality Requirements and Procedures

Commercial standards, product standards and *standard specifications* are some terms for guidelines that identify attributes a product must have to be accepted by consumers and building code agencies.

U.S. commercial standards, such as PS-1 for softwood plywood or PS 20-70 for softwood lumber, plus other like standards for other products are examples of the specification requirements for individual product types. Other codes and standards, such as those promulgated by the Canadian Standards Association (CSA) and the published standards of the Federal Republic of Germany, are used internationally.

Standards are used as reference guides to assess product quality. Each describes specific manufacturing methods and product requirements as well as audit methods and penalties for noncompliance.

The information provided within these standards and codes may be all that is needed for a commodity product. However, those mills that produce a diversified mix of specialty and upgrade items often use a product specification manual.

A *specification manual* is a tabbed notebook containing the mill's complete product specifications for an individual product type. It includes the information obtained from the product standard, and additional specifications that identify particular mill practices are added. Figure 17.1 illustrates a sample page of a multipage guide for softwood plywood.

The *standard operating procedures (SOP) notebook*, a reference guide that is frequently prepared for an individual mill, outlines the manufacturing specifications for each point in the process. Operating practices and machine setup procedures are only a few of the specific instructions included. Figure 17.2. shows a sample page from a typical softwood lumber SOP manual. As with the specification manual, each page is numbered and carries an effective date; appropriate signatures indicate approval.

The SOP manual plus the product standards and the specification manual provide the information that is used as a basis for inspection and testing.

Inspection and Testing

The most effective inspection and testing programs combine the interrelated activities of the product standards inspector, the in-house quality control person, the resin or other vendor and the line supervisor.

The product standards inspector audits the manufacturing process and finished products to ascertain that the minimum product standard is being met.

The in-house inspector, usually the on-shift foreman in smaller mills and the quality supervisor (plus additional technicians of the night shifts) in larger mills, systematically audits the entire process, log yard to boxcar, to ascertain that requirements are being met at least cost.

Status Reporting

An effective quality audit will focus on the vital few from the many. The vital few will change from time to time as a result of product mix changes, process control emphasis and the level of performance at each point in the process. Routine audits should be maintained by on-shift line personnel, while the staff auditor concentrates the lion's share of his effort on pocketbook-type decisions, where a comparatively small amount of effort will result in large savings or value increases.

Figure 17.1. Sample page from a softwood plywood manufacturing specifications notebook. (Source: Baldwin 1981, p. 264)

ABC Building Products

MANUFACTURING SPECIFICATIONS   3/4 ABI

Panel Description & End Use:	A quality sanded panel for exterior and interior use. Primary use for cabinets and wood fixtures in homes.
Veneer Grade & Workmanship:	Six ply panel with A grade face, C grade inner plies, and B grade back as per PS 1-74.

Construction:		Ply #	Veneer Grade	Veneer Thickness
	Face	1	A grade	.166
	Core	2	C grade	.126
Primary	Center	3	C grade	.126
Construction	Core	4	C grade	.126
	Core	5	C grade	.126
	Back	6	B grade	.166

Species:	Southern Yellow Pine
Rough Panel Thickness:	.730 − .762
Finished Panel Thickness:	.750 Tolerance ± 0.16
Glueline:	Exterior Glue GP 31153 mix GP 553 42% resin solids, 28.9% wet mix solids. Monsanto PF 554 mix 3108 45% resin solids, 28.0% wet mix solids.
Recommended Press Schedule:	GP 7 min. @ 285° F 300° F Monsanto 7 min @ 285° F 6½ min. @ 300° F
Special Instructions:	Face veneers come from natural pulled at dryers or from veneer patchers. Panel is to be stamped on end or narrow edge side of each panel. Both face and back receive smooth sand.

APPROVALS

Q.C. Supt. _____

Prod. Supt. _____

Plant Mgr. _____

Effective Date _____

Replaces or Revise _____

Procedure Date _____

Page _____ of _____

Oregon-Atlantic Lumber Co.
          Dept: Sawmill
          Area: Primary breakdown
          Machine center: Headrig

In-process Procedures

A-1. Headrig—Quality Control

1. Ten sample boards from side cuts will be selected at random each shift, more if necessary to solve problems with cuts.
2. The resaw flitches will be checked each shift, front and back measurements taken at least 6" in from end of flitch.
3. Samples will be measured by a caliper rule reading in the 1/32 of an inch up to 6".
4. Results will be recorded on Quality Control Data Work Sheets and will include target size.
5. Measurements will be made at the top and bottom edge as it comes from the saw, starting 18" from the head and measuring three places, minimum, the length of the pieces.
6. Results of measurements will be recorded on Quality Control Data Work Sheets and include target size set on each length station measured. The number of 1/32" below target size are negative and shall have a minus (−) sign. Enter top measurements above the diagonal line and bottom measurements below.
7. Record under "Remarks," if required, unusual log conditions, condition of saw, sharp or dull, relief or substitute sawyers, etc. Enter guide height if excessive in proper column (inches above cut), and compliance of operating procedures.

    Note: All operating and maintenance manuals are included as Appendix B. See individual machines or stations.

A-2. Headrig—Operating Procedures (Related to Quality Control)

1. Logs should be loaded (all that are possible) with the highest grade side of each log being sawed first.
2. All knees should be used such that the length of the log allows for balance and will hold securely while being sawed.
3. All cuts shall be made to conform to cutting orders.
4. For the greatest realization and grade recovery, light slabbing should be the rule, preferably 4"×8' but not over 6" for the first cut.
5. Use the maximum taper that is needed to eliminate short cuts. Work for the largest percentage possible in log cuts. Take the taper out after the rough has been removed.
6. All logs should be manufactured by taking all clear from one side before turning to next side, thereby eliminating much of the log turning and dog damage to the lumber. After the clear has been turned and cut off in this way, the heart should be boxed in the center of the remaining cant or flitch. Rough logs should be turned to remove the larger outside knots first to eliminate spike knots and retain the maximum value of the log.
7. The sizes shall conform to the posted target tolerance limits plus or minus 1/16" from actual target size.

continued on next page

Figure 17.2. Sample page from a softwood lumber SOP manual.

8. All *setworks* shall be *checked twice daily* by using multiple sets. Then the boards run dial will readily show any discrepancy in the individual sets, or setworks that is erratic.
9. Sawyers shall make these checks and report any difficulty to their foreman immediately for correction.
10. All saws shall cut a quality product.
11. The gullet of the saw shall be no more than 3/4'' from the rim of the band mill wheel.
12. If a saw is damaged while in use or is not cutting properly, notify the filers.
13. The guides must be checked each half shift and be reset if necessary to assure that the saw is in a true parallel line with the vee rail.
14. It shall be the responsibility of the *sawyers* to ascertain that the strain points are kept free of sawdust and pitch buildup by blowing them off when changing saws; that the strain arm is in position prescribed by the head filer; and that the saws are properly balanced on the wheels.
15. Use target size sets to eliminate shim waste and extra lines.

B. Flitches—Quality Control
Four sample pieces shall be measured each half shift and recorded. Measurements to be taken on top of piece with five measurements starting six inches from either end. Ends will be checked for square and thickness.

C. Headrig Cants—Quality Control
Four pieces shall be measured twice daily and recorded on work sheet; one measurement from each end. Read measurements in 1/16''.

Approvals

Q.C. supt. _____

Dept. supt. _____

Produ. supt. _____

Effective date _____

Replaces or revises
procedure dated: _____

Page _____ of _____

Figure 17.2. *Continued*

The vital few in the lumber process include green lumber size checks, kiln-schedule and moisture-content checks, and skip checks at the planer. Final product checks are twofold. The grade will be audited at timely intervals by a check grader to ascertain that the grade standard is being met. Secondly, a lumber degrade evaluation will be made to determine the cause of the degrade, whether natural defects, kiln defects, manufacturing deficiencies such as excessive skip due to poor sizes, or misgrading. The objective is to determine the preventable cause and to move quickly to correct the deficiency.

Plywood's vital few include peel quality (smoothness of the face, tightness of the back and uniformity of the peel thickness), veneer drying (moisture content level and uniformity of that level), gluing results (such as percent wood failure and failing panels) and downstream regrading in the warehouse. On-grade checks and downgrade analysis provide the overview for the entire manufacturing process.

Particle size, alignment, moisture content, resin and additive addition, plus monitoring of board properties such as mat/board density and thickness are a few of the many checks in the reconstituted board process.

Reporting and interpretation of the results rely on two basic statistical concepts: accuracy and consistency. *Accuracy* demonstrates the ability of the individual employee or the capability of a machine to attain a preset standard. Accuracy is measured by the arithmetic mean ($\overline{X}$) or identified by the mode.

Consistency, the second concept, is usually measured by the difference between the highest and the lowest value. This measure of dispersion is called range (R). The calculation of standard deviation (σ) is also a good measure of consistency; this concept is used frequently with computer-controlled readouts, which are commonly used in conjunction with the sophisticated process control devices. Figure 17.3 is an example of manual use of the two statistical concepts, accuracy and consistency, in a typical within-product evaluation.

Figure 17.3. Calculation of within and between-board variation.

Record sheet for $\overline{X}$ and R chart

Material or machine center __GANG EDGER__

Characteristic measured __WITHIN AND BETWEEN — BOARD VARIATION__

Plant __TOLEDO__ Area __SOUTHERN OREGON__

Unit of measurement __.001__ Recorded by __C + B__

Series no.	Date	Shift	A	B	C	D	E	X̄	R	Record of inspection
1	2/21	1	1.685	1.700	1.690	1.635	1.683	1.679	.065	
2	7:15	1	1.686	1.645	1.710	1.650	1.684	1.675	.065	
3	A.M.	1	1.725	1.750	1.675	1.635	1.591	1.675	.159	Side board
4		1	1.676	1.690	1.625	1.650	1.671	1.662	.065	
5	5:05	2	1.676	1.625	1.735	1.590	1.725	1.670	.145	Smoking due
6	P.M.	2	1.736	1.543	1.650	1.650	1.650	1.650	.173	to overfeeding
7		2	1.685	1.710	1.680	1.637	1.685	1.679	.073	
8		2	1.685	1.655	1.685	1.687	1.685	1.679	.032	
9	2/22	1	1.684	1.685	1.710	1.624	1.680	1.677	.086	
10		1	1.685	1.680	1.637	1.685	1.688	1.675	.051	
11		1	1.680	1.685	1.625	1.685	1.690	1.673	.065	
12		2	1.670	1.736	1.690	1.676	1.675	1.689	.066	
13		2	1.675	1.625	1.685	1.690	1.686	1.672	.065	
14		2	1.736	1.690	1.736	1.684	1.685	1.706	.052	
15	2/23	1	1.686	1.690	1.685	1.625	1.690	1.675	.065	
16		1	1.685	1.687	1.625	1.685	1.590	1.654	.097	
17		1	1.690	1.685	1.737	1.690	1.656	1.690	.087	
18		1	1.612	1.625	1.735	1.685	1.710	1.689	.110	
19		2	1.610	1.687	1.625	1.732	1.635	1.658	.122	
20		2	1.637	1.685	1.738	1.687	1.638	1.677	.101	
	Totals							1.675	.087	

Target — 1.675"

A series of 20 sample boards are measured at the outfeed of the gang edger. Each board is measured at five equidistant points starting about 12 in. from each end. The five measurements along the board length are tallied and then averaged. The resulting statistics indicate the within-board (or product) accuracy.

The second statistic in Figure 17.3 is the range. This number represents the difference between the highest and lowest value. It thus documents the consistency of processing at this machine. The resulting information can be quickly evaluated to determine if further action is warranted or if the process is in control, that is, within a plus or minus tolerance to a target attribute.

Figure 17.4 is another documenting and reporting format, which is intended to quickly identify the between-product accuracy and consistency. The chart indicates the sawing accuracy of a double-cut headrig, a bull edger and a downstream double-cut bandmill.

For example, the forward pass of the double-cut headrig results in 80.6% accuracy for the 186 samples measured, based on an accuracy tolerance of $\pm 1/16$ in.

Figure 17.4. Target size check.

Check point		Headrig		Bull edger		Double-cut	
Target sizes		Forward	Backward	East	West	Forward	Backward
Saw kerf							
Inches oversize	+3/8						
	+11/32						
	+5/16						
	+9/32						
	+1/4					1	
	+7/32		|||| (8)				
	+3/16			(1)				
	+5/32	||| (2)					
	+1/8	(17)	(10)			(1)		
	+3/32	(16)	(10)	||| (3)			
	+1/16		|||| ||| (8)	|||| ||| (8)	(13)	||| (3)	|||| (5)
	+1/32			(21)	(35)	(45)	(24)
Target size = 0		(10)/(40)	(10)/(40)	(34)	(30)	(32)	(39)
Inches undersize	−1/32		(12)	|||| (5)	||||	(6)		
	−1/16		|||| (5)	1 (1)	||| (3)		
	−3/32		1 (1)				
	−1/8						
	−5/32						
	−3/16						
	−7/32						
	−1/4						
	−9/32						
	−5/16						
	−11/32						
	−3/8						
Tot. measured		186	193	72	88	80	68
In / **Out**		36	26	3	1	0	0
% In tolerance		80.6%	86.5%	95.8%	98.8	100%	100%
% Out tolerance							

The consistency to target size varies from 0 to +5/32 in. Then note that consistency ranges from a −3/32 from target to a high of +7/32 from the target size on the backward pass.

Figure 17.5 is a format used to audit peel thickness. It illustrates the results of measuring 154 pieces of green 1/6 (.165) veneer fishtails. This crossband veneer,

Figure 17.5. Veneer thickness range.

Nominal thickness-type 1/6 FISHTAIL

Source STORAGE **Date** 4-28-84

Time

Thickness 1/10-1/8	1/6-3/16	Tally (5 · 10 · 15 · 20 · 25)	Count	Value	Notes
.085	.140	xx xx	4	.56	
.086	.141	x x	2	.282	
.087	.142	x x	2	.282	**Thin veneer**
.088	.143				(less than
.089	.144	x	1	.144	.154)
.090	.145				
.091	.146	x	1	.146	Pieces %
.092	.147	x x x x	4	.588	30 19.5
.093	.148				
.094	.149	x	1	.149	
.095	.150	x x x x x	5	.75	
.096	.151	x x x x	4	.604	
.097	.152	x x x x	4	.608	
.098	.153	x x	2	.306	
.099	.154	x x x x x	5	.77	
.100	.155		7	1.085	
.101	.156		5	.78	
.102	.157		4	.628	
.103	.158		7	1.166	**Average**
.104	.159	x	1	.159	.162
.105	.160		11	1.76	
.106	.161		2	.322	Pieces %
.107	.162	Ave. .162	7	1.134	103 66.9
.108	.163		10	1.63	
.109	.164		8	1.312	
.110	.165		11	1.815	
.111	.166		3	.498	
.112	.167		4	.668	
.113	.168		2	.336	
.114	.169		8	1.352	
.115	.170		8	1.36	
.116	.171	x	1	.171	
.117	.172		2	.344	**Thick veneer**
.118	.173		2	.346	(more than
.119	.174		2	.348	.170)
.120	.175		2	.35	
.121	.176				Pieces %
.122	.177				21 13.6
.123	.178		3	.534	
.124	.179				
.125	.180				
.126	.181				
.127	.182				
.128	.183		2	.366	
.129	.184		2	.368	**Veneer quality**
.130	.185		3	.555	% ±5%
.131	.186		1	.186	of average
.132	.187				
.133	.188		1	.188	70%——Good
.134	.189				
.135	.190				60%—— Fair
.136	.191				
.137	.192				50%—— Poor
.138	.193				
.139	.194				
.140	.195				
.141	.196				
.142	.197				
.143	.198				
.144	.199				
.145	.200		154	24.892	

Comments

On-shift personnel should conduct routine audits in order to catch problems before they multiply.

an acid test for a lathe setup, shows that only 66.9% of the samples measured fall within ±5% of the average. The thickness varies and is relatively inconsistent; the thickest piece is .188 and the thinnest is .140 in.

The three figures illustrate formats that can be useful tools in determining the results achieved in meeting size targets at an individual machine center with a particular manufacturing capability. Each is a useful visual aid in determining current performance.

Data Analysis and Corrective Action Procedures

The cumulative data are observed, documented and then communicated by the auditor, with subsequent and frequent follow-up by the assigned individual. Prompt corrective action is the activity that determines whether a quality control program is effective.

Timely action and follow-up indicate management's commitment to achieving results. This commitment is common to all successful quality and process control programs. The corrective action should include a formalized procedure to resolve the problem.

For example, when an out-of-control situation occurs, the on-shift supervisor should have the first real opportunity to correct the problem; the situation is then escalated if correction does not occur within an agreed-upon time interval. The step-by-step escalation procedure continues until the problem is resolved.

If it takes two or more steps to correct the problem, this is a good indication that the organization is not functioning as smoothly as it should. The problem then becomes two problems: the identified problem plus the people problem that let it occur and/or continue.

The manager then resolves the reported quality problem and proceeds to correct the overriding people problem. The problem-solving activities of the manager also include encouraging input from both salaried and hourly employees. One such program is the quality circle.

THE QUALITY CIRCLE AND PROBLEM SOLVING

Quality circles or involvement programs, fairly recent innovations in the forest products industry, can be successfully used to add to or complement the formal mill program described above. There are many program variations that will garner the ideas of hourly and salaried employees. The success of each will hinge on the leadership provided, the recognition given for thoughtful ideas and the ability of the workgroup to overcome resistance to change.

These programs are composed of one or more small groups of employees who, on company time, meet regularly and voluntarily to discuss quality problems and then formulate solutions. The quality circle concept is based on the idea that each employee is capable of contributing far more than time and repetitive activity. It seeks to involve employees in matters that affect them, matters about which the workgroup has a unique opportunity to contribute ideas and solutions.

"The theory is that the workers know their jobs better than anyone else, and, given a chance, will be creative and self-motivated" (Knight 1982, p. 6).

Quality circles combined with a formalized auditing and follow-up program provide a full coverage quality control program. Such a program will prevent mistakes, and better yet, will make a continuing contribution to the business.

REFERENCES

Baldwin, R. F. 1972. Ways to Increase Sawmill Overrun. *Wood & Wood Products* 77(8): 110B–110D (August).

———. 1981. *Plywood Manufacturing Practices*. Rev. 2d ed., Ch. 17. San Francisco: Miller Freeman Publications.

Brown, T. D., ed. 1982. *Quality Control in Lumber Manufacturing*. San Francisco: Miller Freeman Publications.

Crosby, P. B. 1979. *Quality Is Free*. New York: McGraw-Hill.

Fransworth, J. 1980. Use of Sawmill Quality Control Data in Setting Equipment Specifications and Acceptance. Speech given at Forest Products Research Society, Pacific Northwest Spring Meeting, May 1, 1980.

Grant, E. L. 1964. *Statistical Quality Control*. New York: McGraw-Hill.

Knight, D. K. 1982. QC's: Worth A Diligent Try. *Timber Processing* 7(11):6 (November).

Maloney, T. M. 1977. *Modern Particleboard & Dry-Process Fiberboard Manufacturing*. San Francisco: Miller Freeman Publications.

Maloney, T. 1983. Quality Assurance, Growing Manufacturing Priority. *Plywood & Panel World* 24(1):10 (January).

Moslemi, A. A. 1974. *Particleboard. Volume 2: Technology*, pp. 162–163. Carbondale, Ill.: Southern Illinois University Press.

Eighteen
Managing Maintenance and Purchasing

A maintenance program means different things to different people. To some it is little more than the outmoded concept "Keep a little grease on it and then run it until it quits." Most mill managers have discovered the losses inherent in such a program and have sought something more.

PROGRAM OBJECTIVES AND FEATURES

An effective plant maintenance program has a number of objectives:

- To prevent breakdowns and minimize the frequency and extent of lost time
- To maintain plant equipment at peak efficiency to prevent yield losses and product degrade
- To utilize the equipment within stated safety requirements
- To obtain the optimum level of maintenance at the optimum cost, i.e., to get the "biggest bang for the buck"

Whatever the converting process, whether manufacturing a lumber, plywood, board or another softwood-based product, maintenance programs that meet these objectives will share several features. As a rule of thumb the larger the operation or the more complicated the manufacturing process, the more organized and detailed the program must be to be effective. The following features are common to all effective programs:

- A clean mill with low downtime and high worker productivity
- A strong manager with maintenance appreciation who exerts forceful local control
- Effective communication and cooperation between production and maintenance personnel

OUTLINE OF A MAINTENANCE PROGRAM

The following activities make up a total maintenance program: staffing, planning and scheduling, record keeping and cost tracking, maintaining a total lubrication program, safety and loss prevention, purchasing, and maintenance prevention.

The Organization and Its People

Good maintenance performance is the result of a good organization. The lines of responsibility, the authority to take action and the duties assigned are spelled out within the framework of the organization.

The more effective maintenance organizations usually report to the unit manager. Technical strength and supportive maintenance appreciation are key attributes for this manager. This unit manager, whether titled a department superintendent, plant manager, operations manager or something else, will seek balance, a balance that will ensure teamwork and good communication between production and maintenance personnel.

The maintenance department will be organized with well-defined lines of authority. Within a multiproduct complex, these lines of authority will originate within the unit or the product organization rather than taking the form of a centralized organization reporting to the complex engineer. Unit maintenance provides narrower accountability, tighter controls and easier communication.

A central organization is utilized only to supply specialized services such as fire protection, mobile equipment maintenance, machine shop services and other common functions. A company with geographically dispersed operations will sometimes use a specialized task group to enhance the area maintenance concept. This group will provide engineering support services plus specialized troubleshooting, such as boiler or electronic technical service.

Training and upgrading are continuing tasks. The success of this effort will be determined by the individual and collective performance of each hourly maintenance employee as measured by the quantity and quality of the work, whether whistle chasing, benchwork in the shop, or outside-project activities. The supervisor who assigns and delegates must also spark the educating and upgrading process for each individual maintenance person.

An effective training program will document and grade the various skill levels and then assess the needs of each individual team member. Figure 18.1 is an example of a skill audit matrix. As shown, the millwright position is graded from skill level D through A+ with a further advancement to leadman if the leadership capability of the individual matches the needs of the organization. The electrician's job is evaluated in a similar fashion. A multicraft maintenance person may be at different levels within each of the two skills areas. More specialized skills are handled as exceptions, which are rated on an equally specialized basis.

The maintenance supervisor, using this or a similar matrix, then objectively assesses the needs of each employee for training. A progressive program will have skill and knowledge requirements at each level. The employee is usually allowed to set goals and to set the pace for achieving those goals within limits. On-the-job training (such as cross-training), locally available technical courses and vendor schools are some of the methods used.

Millwright	Electrician
D	**D**
Mechanical aptitude; willingness to learn; adaptable; good judgment; ability to work safely; basic knowledge of hand tools and portable equipment; serve as a helper.	Electrical aptitude; willingness to learn and follow instructions; good judgment; safe work habits; show initiative; basic knowledge of tools and equipment; serve as a helper.
C	**C**
Basic knowledge of fundamentals of millwrighting, including welding, burning, measuring devices, lubrication. Basic machine repair, maintenance and troubleshooting. Basic knowledge of oils and greases; some shop math; able to work safely.	Basic knowledge of fundamentals of electricity. Knowledge of electrical tools, ability to thread conduit, hang lighting fixtures; assist electrician as needed.
B	**B**
All of *C* plus more advanced skill level with troubleshooting capability. Show initiative. Floor man on routine shift with ability to handle routine problems. Basic knowledge instrumentation repair.	Ability to read blueprints, make splices, hook up wires from blueprints; basic knowledge of electrical troubleshooting procedures. Knowledge of electrical meters. Basic knowledge of and ability to test motors and generators; ability to change out motors.
A	**A**
All of *B* plus advanced skill levels in welding, couplings, pumps, shafts, etc. Able to set bearing tolerances, high level of burning skills. Advanced troubleshooting ability. Advanced knowledge of blueprint reading and sketching. Basic knowledge of pipefitting, working knowledge of temperature and pressure controls. Ability to plan materials, work time and equipment necessary to complete job; utilize people effectively. Ability to install air control of valves and fittings.	All of *B* plus advanced skills in reading prints; some knowledge of solid state equipment. Generally able to wire panels and electronic systems; advanced troubleshooting capability. Able to handle high voltage distribution problems; install electronic automatic control valves. Knowledge of repair and maintenance of motors and generators; knowledge of motor settings, drives, coupling devices, manual and automatic controls including magnetic and solid state. Basic knowledge of environmental metering devices. Ability to lead others. Basic knowledge of local and state electrical codes.
A+	**A+**
All of *A* plus basic knowledge of electricity; exceptional ability in welding and troubleshooting. Leadership ability; ability to set priorities.	All of *A* plus complete knowledge to design and install motor circuits as required; ability to operate electrical shop, take charge and make decisions; ability to lead people; experienced troubleshooting abilities, identify problems quickly and take action, basic knowledge of electronics, DC circuiting, drives and controllers. Ability to organize and set priorities.
Leadman	**Leadman**
All of *A+* plus ability to organize many projects; coordinate the various trades. High level of leadership ability. Full understanding and appreciation of management goals. Ability to evaluate people.	All of *A+* plus ability to organize work assignments for crews; line up work. Ability to take charge in foreman's absence. Generally high level of leadership ability. Full understanding and appreciation of management goals. Ability to evaluate people.

Figure 18.1. Maintenance skills audit matrix.

A multicraft organization is more effective than a single staff of specialized millwrights and electricians. However, a multicraft organization requires more selective hiring, a comprehensive training effort and a continuing commitment if it is to obtain its intended benefits of tighter crewing and flexible job assignment.

A multicraft organization recognizes that a crew member is a maintenance person first and a craft worker second. Although certain individuals in a crew will have specialized skills, the primary task of each is to keep the mill running and the downtime minimized.

Planning and Scheduling

The planning and scheduling function shifts the maintenance effort from an inefficient breakdown-and-fix approach to a scheduled preventive program. The first step in planning and scheduling is the control of time and tasks.

The *work order* assists the planner in assigning man-hours and resources to the jobs being scheduled. Figure 18.2 is an illustration of a two-part work order frequently used. This work order provides an organized method for describing the problem, identifying the location, prioritizing the need, and designating the completion date.

The two-part form provides a completed top copy to the originator as a reminder; the card-weight back copy is used as a planning and scheduling tool. When the work is completed, the work order becomes a source document for the machine or equipment record.

The work order is an excellent communication and follow-up device that includes a description of the work completed, the parts and materials used, the job costs, and the man-hours expended; it also identifies the maintenance person who completed the job. It is a scheduling and follow-up tool that aids in solving maintenance problems or implementing changes in the process. In addition the work order assists communication and understanding between production and maintenance personnel.

A *preventive maintenance checklist* is another aid used to schedule man-hours and resources. It is prepared for each machine or machine cluster in a department. A maintenance person or operator is then assigned to periodically audit each item listed; delegating this procedure to a specific individual aids accountability.

Progressive wear or other visible conditions are evaluated during periodic, routine checks. These checks will turn up a list of items to be replaced or repaired; the list is then prioritized and scheduled. Replacing a frayed flatbelt lacing between shifts or changing out a noisy gearbox, as examples, will frequently prevent costly production delays.

Figure 18.3 is an example of a checklist used for a specific panel hot press charger in a West Coast mill. Downtime was reduced 30% and more at this machine center after the checklist was put into use. This checklist allows the supervisor and maintenance planner the opportunity to pick repair times or figure out ways to avoid costly random production delays.

Work orders, production and maintenance checklists, plus oiler-supplied rough notes identify the lion's share of the maintenance work to be accomplished. The planner, usually the maintenance foreman or superintendent, evaluates the input of

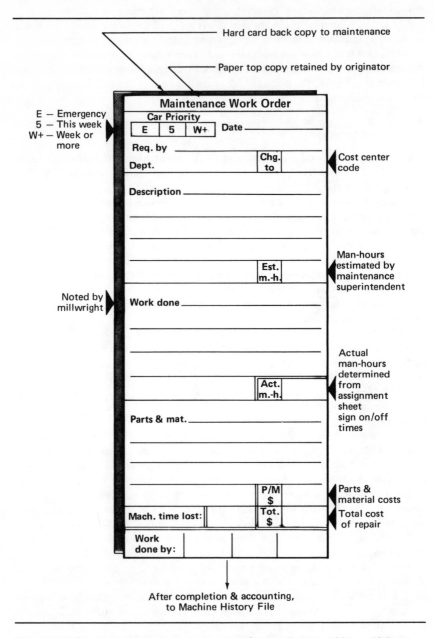

Hard card back copy to maintenance

Paper top copy retained by originator

E — Emergency
5 — This week
W+ — Week or
more

Noted by
millwright

Maintenance Work Order

Car Priority

| E | 5 | W+ | Date |

Req. by

Dept.   Chg. to

Description

Est. m.-h.

Work done

Act. m.-h.

Parts & mat.

P/M $

Mach. time lost: Tot. $

Work done by:

Cost center
code

Man-hours
estimated by
maintenance
superintendent

Actual
man-hours
determined
from
assignment
sheet
sign on/off
times

Parts &
material costs

Total cost
of repair

After completion & accounting,
to Machine History File

Figure 18.2. Standard maintenance work order. (Source: Baldwin 1981, p. 284)

the other supervisors and hourly employees. He then plans and develops the maintenance schedule as follows:

- Identifies the project and documents the agreed-upon scope.
- Establishes a priority for each task. In the simpler systems, tasks can often be classified as musts, needs, and wants. Other systems will use numerical scales, such as a rating of 1 through 10, to prioritize tasks.

Figure 18.3. Weekly preventive maintenance checklist. (Source: Baldwin 1981, p. 280)

WEEKLY PM CHECKLIST

When you check the items listed below, put your initial in the space provided.

Charger Pans

_____ 1. Straighten or replace any damaged charger pans.

_____ 2. Straighten any bent nosebars.

_____ 3. Check saddle bolts for tightness.

_____ 4. Check brass nosebars for damage and repair as required.

_____ 5. Inspect and repair nosebars on hot press for damage and proper installation.

_____ 6. Bring charger to press, do not charge pans into press, check each pan for 1/32" min. to 1/16" max. clearance at press platen.

Pusher Bars

_____ 1. Return charger to load position.

_____ 2. Repair or replace any bent or missing pusher bars.

_____ 3. Inspect 50B15 sprockets on pusher bar drive shaft.

_____ 4. Inspect split sprockets, bushings and bearings on pusher bar drive shaft.

_____ 5. Tighten all chain deflector bracket bolts and replace any bent chain tighteners.

_____ 6. Tighten all pusher bar chains and insure washers are in place under chain tighteners.

_____ 7. Inspect all drive couplings and bearings on top of unloader.

_____ 8. Time all pusher bars.

_____ 9. Operate pusher bars to insure proper operation.

_____ 10. Recheck pusher bar time.

Lubrication

_____ 1. Grease all rack and pinion gears on charger.

_____ 2. Grease pusher bar drive bushings.

_____ 3. Check all gearbox levels.

_____ 4. Oil all pusher bar chains and chain tighteners, idlers with LPS #3.

Foreman's name: _____ Date: _____

Comments:

- Designates resources (man-hours, materials and parts, and other available resources) and assigns each to the job.
- Communicates the schedule and results expected to the doer and the requester. Keeps each informed before, during and subsequent to the project.
- Organizes the work prior to the start of the shift or job. This includes reviewing the work method and the instructions provided and preparing the work site whenever possible. Needed equipment, parts and materials and other resources are set up at the jobsite.
- Tracks and follows up on the work in progress. Compares the results to planned or calculated should-be times and resource use.

Using these guidelines, the planner then lines out the work based on an organized and prioritized work list. He identifies specific jobs to be done at specific times using specific resources. Timely feedback is given to the manager or department head.

Record Keeping and Cost Tracking

The complexity of maintenance record preparation and reporting will vary with the complexity of the operation. Each operation will have some form of the following:

Work lists and work orders
Downtime reporting systems by machine center
Material and parts use information
A maintenance history for each machine center
Files of manufacturer information and prints
Electric motor and gearbox records
Maintenance checklists
Lubrication records

These records provide the basis for ongoing tracking. Tracking is useful when the monthly statement shows that less than optimum results are being achieved; record keeping is also useful as a control device to audit ongoing maintenance.

Mobile equipment cost and service histories are the most common applications for cost tracking and accounting. Many companies use the information to isolate unfavorable trends, such as operator abuse or misuse. This type of cost tracking and accounting is also increasingly being used in the mill at individual cost centers. The maintenance records combined with cost accounting can identify excessive costs and pinpoint possible solutions.

Total Lubrication Program

The lubrication program is the heart of the maintenance effort. The program has a twofold purpose: (1) to ensure that each lubrication point is serviced at the just-right time with the just-right quality and quantity of lubricant and (2) to serve as a vehicle for a systematic inspection of the equipment and its components.

A program is initiated by conducting a comprehensive audit at each machine. This audit, usually performed by a reliable vendor, identifies the type, quantity, and frequency of lubrication needed. The needs established by the audit are then translated into a lubrication schedule.

The lubrication schedule identifies the lubrication points by daily, weekly and/or monthly requirements. The quarterly, semiannual and annual portions of the lubrication list are frequently computerized to provide a timely reminder.

The plant oiler has a unique position. His normal activities take him to remote onsite locations that are seldom visited by other personnel. The deteriorating operating conditions of gearboxes, bearings and hydraulic systems can often be identified, tracked and repaired before unscheduled downtime occurs.

"Oilers were the first line of inspection," cited a 1976 unpublished report (Weyerhaeuser 1976). Many plants provide the oiler with a pad and pencil to document problems for prompt follow-up; a few provide a miniature tape recorder for the same purpose.

In addition to properly lubricating the equipment at timely intervals and documenting problem areas, the oiler is increasingly being asked to conserve petroleum products. One California mill initiated a conservation program in conjunction with the plant's suppliers. Together they developed a program that includes:

1. Avoiding spills, waste and contamination by controlling inventory and use
2. Controlling oil leaks by inspection and follow-up
3. Determining hydraulic oil life by laboratory analysis
4. Determining engine oil life by laboratory analysis
5. Extending oil life through purification and reclamation of used oil

Conservation programs are gaining wide acceptance as each mill seeks to reduce reliance on increasingly more expensive petroleum-based products.

Safety and Loss Prevention

The organization and its people are important; also important is the environment within which each individual performs. Housekeeping, safe work practices and loss prevention activities shape the manufacturing environment.

Abnormally short belt life, worn conveyor chain, rapidly deteriorating equipment housing or structures, broken or out-of-place equipment guards and frequent and extended downtime are usually symptoms of sloppy housekeeping.

Clean up, find, and keep clean are action steps in a sequence of activities by which the operator zeros in on debris, loose parts, and trash. This is an important indication to the maintenance and operating personnel that efficient operation and quality are important. No less important is safety.

Actual and potential hazards are recognized before the repair or fix-up job begins. Power systems are locked out in the immediate area and failsafe devices are activated. OSHA standards are applied as minimum standards, with other rules, policies and common-sense ideas implemented as appropriate. The tag-out procedure should be used at all times; the employee who places the tag should remove the same.

The job is never completed until the guard is in place and the maintenance materials are cleaned up. Discarded bearings and metal parts such as bolts are frequently responsible for a seemingly endless sequence of fix and repair activities as the discarded parts are swept into conveyors or become entangled in sprockets and pulleys . . . and blow up or seriously damage chippers, hogs and feeders.

The Purchasing Function

An effective purchasing program will order, stock and control parts and materials so that the just-right replacement part or material is available at the right time. A fiberboard mill incident is an example of what not to do.

A breakdown occurred on the evening shift and the special bearing needed to fix it was unavailable, although several had been purchased. The mill resumed operation after a millwright traveled home and sorted through another tool chest for a like bearing. The part had been "salted away" in the event of just such an occurrence but had inadvertently been placed in another toolbox.

Inventories of spare parts should be secured in a controlled location with some means to identify use and the restocking schedule. These controls will minimize downtime and provide the opportunity for the buyer to shop and compare. A southern mill exemplifies the benefits of one such system.

An equipment vendor had been supplying a special fuse for a green veneer tray system control. Silver, a prime component of the fuse, rose in price and the fuse tripled in cost; a $60 cost became $180 overnight. Because of the lead time provided by a cataloged and closely controlled inventory, the buyer researched and found the same product offered by another vendor. The $180 cost was reduced to $54 without sacrificing quality or dependability.

Buying for value is an ongoing process that incorporates the search-and-find process but also considers substitution, standardization, simplification and even elimination as part of the total effort to reduce costs.

A *critical spares program* (CSP) complements the total purchasing and maintenance effort. Critical spares, defined as the vital few specialized parts (see Figure 18.4), are handled as exception items. An exception item may be a special one-of-a-kind gearbox for the fuel chain; it may also be a special drive unit for the forming line. Major downtime of shifts or days would occur if backup units were not immediately available when equipment failure occurred. The CSP program is an insurance policy that will prevent unnecessary downtime. The CSP implementation steps are as follows:

1. Form a Critical Spares Committee at the location. This committee should consist of the operations/plant manager, the local purchasing representative, the maintenance manager/superintendent, and others as required.
2. Review each service or machine center and identify the critical few parts as defined in Figure 18.4.
3. The operations/plant manager, local purchasing representative, and maintenance manager/superintendent further screen the list to determine if all entries are indeed critical spares, based on possible contingencies and the probability of occurrence. For example:
    a. A part may be critical but the probability of its unexpectedly becoming defective is extremely low. Therefore, rather than stocking the part, a periodic preventive maintenance audit is scheduled.
    b. A part may be critical for a machine but the machine may not be critical

to the operation. Therefore, the required part may be stocked only as part of the normal inventory.

4. The local purchasing manager evaluates the list to determine the following:
   a. Is it necessary to stock the part in-house or can an agreement be made with a reliable local vendor to retain the part in stock?
   b. Do other reliable sources (mills with like equipment) have the part readily available? Can an agreement be made with them to stock the part?
   c. What are the acceptable stocking limits for the part (minimum and maximum quantities)?
5. The operations manager and the purchasing representative then determine the expenditures required and the timing of the expenditures.
6. The local purchasing representative then implements the plan:
   a. An approved list is typed up and put in a red looseleaf binder. Each inventory sheet is covered with plastic so that a grease pencil can be used to keep a dated running inventory on each part. A minimum and maximum inventory is identified for each part; the minimum level is usually the reorder point.
   b. The incoming critical spare parts are put in a central location, with department identification marked in large letters that are easy for all maintenance personnel to see.
   c. A tag (color-coded by machine or department) is attached to each part; the item purchase order description is on the tag.

Figure 18.4. Critical spare parts: a definition. (Source: Baldwin 1981, p. 282)

A part or component which would cause a substantial shutdown of all or a significant part of a primary production line if the existing part were to become defective is a critical spare part.

Critical Spares Are:
1. Based on the concept that 20% of the machine components will create 80% of the downtime. The vital few of the many.
2. Based on 1, above, the 20% that are truly critical are:
   a. Not easily repairable within an acceptable time frame.
   b. Sometimes a custom made or proprietary part that requires long lead time.
   c. The part may be only available at a distant location and is not usually stocked locally.
   d. Usually a component for a critical machine that does not have a backup system.

Critical Spares Are Not:
1. Normal replacement parts that are readily available; these parts are handled as part of normal warehouse inventory.
2. Usually not components for equipment that can remain idle until the part is received without impacting mill operations.

d. Each time a part is used from inventory, the tag is removed by the maintenance person or the storekeeper and the book is checked to see if the reorder point has been reached. If it has, the part is ordered immediately.

e. The critical spare parts in stock are physically inventoried at least three times each week to reconcile the perpetual record with the physical.

f. Critical spares stocked by a vendor or by other mills are periodically checked to determine contingency availability.

7. The list is reviewed periodically by the Critical Spare Parts Committee and revised as needed. Usually a quarterly review is adequate; a new capital item would require an immediate review.

A CSP is part of an overall program that seeks to minimize the stores inventory while ensuring the timely availability of needed parts.

Maintenance Prevention

A maintenance prevention program includes all the elements of the maintenance program; it also includes a questioning attitude, a conscious effort to design away from maintenance, and workplace simplification.

Questioning Attitude. The supervisor, the maintenance person and the machine operator all need a questioning attitude to seek the cause of each downtime occurrence, to prevent future occurrences and to minimize the effects of each.

What happened? What are the facts? How long was the equipment down? What was done to minimize the impact of the downtime? What will be done to prevent a recurrence? The theme is "Everybody is part of the maintenance program, but only a few specialized individuals carry tools." The questioning attitude usually leads to the next maintenance prevention concept: the conscious effort to design away from maintenance.

Preventive Design. Cited a maintenance man, "I contend, then, that the best preventive maintenance dollar is spent during the design stage by close attention to detail and good system design."

Self-cleaning tail pulleys, nonlubricated bearings, special pumps and gearboxes, equipment standardization and the right selection of process control sensors are only a few of the many equipment and component types and designs that can be fitted into the initial construction or retrofitted at a later date (usually at a higher cost). Too often a project engineer will take shortcuts that result in lower initial costs but cause significantly higher operating costs.

The task is to recognize the tradeoffs in the design stage and then incorporate time-tested maintenance prevention ideas into the project. No less important is workplace simplification, which begins with startup and continues through the life of the facility.

Workplace Simplification. The third maintenance prevention element, workplace simplification, involves examining each machine center or area to eliminate unneeded items—those that do little if anything and can be removed from the mill. Items of this sort generally fall into two categories:

• Mechanical items that add little to the process

• Unused and obsolete equipment that is just in the way of the normal maintenance and production activities

Separating the trees from the forest is the task of the operator and his maintenance staff. A manager has to become increasingly aware that maintenance includes all aspects of the operation. Unit costs, operating expenses, equipment utilization, product yield, downtime hours, personnel safety, housekeeping and the general appearance of the plant are important to maintaining a competitive and profitable operation. A wise operator will design his maintenance program carefully and devote the attention and time required to achieve the desired operating results.

REFERENCES

Baldwin, R. F. 1981. *Plywood Manufacturing Practices*. Rev. 2d ed. San Francisco: Miller Freeman Publications.

Buell, D. A. 1973. Management of the Maintenance Function for Manufacturing Plants. Delivered before the Simpson Timber Co. Plywood & Door Annual Meeting, Eureka, Calif., February 20, 1973.

Coleman, M., ed. 1980. *Maintenance Methods for the Pulp and Paper Industry: A Collection of Articles from Pulp & Paper*. San Francisco: Miller Freeman Publications.

LaBelle, B. 1973. Setting the Mill to Minimize Maintenance. *Modern Sawmill Techniques*, Vol. 1, p. 288. Proceedings of the First Sawmill Clinic, Portland, Oreg., February 1973. San Francisco: Miller Freeman Publications.

Weyerhaeuser Co. 1976. Maintenance Effectiveness Study—Summary. Tacoma, Wash.

Nineteen
Productivity and Work Analysis

One way to increase profitability is to increase productivity. Better methods can extract ever greater efficiency from existing resources such as people, machines and raw material. *Industrial engineering, methods analysis, time and motion study*, and *production engineering* are a number of more formal terms used to describe the improvement process.

Various management analysis tools are used to determine and implement improvements. The current operation is assessed; the tasks are broken down into manageable segments, the operation is quantified, the resulting facts are analyzed and the solution is determined using systematic methods.

The following stair-step of activities is the route to reaching cost-saving goals:

- Identify a cost center or process where the costs are high and the results less than optimum. Investigate and then develop an awareness and sensitivity to the possibilities.
- Select the study method. The properly selected study method will not only identify the facts, but it will also highlight inefficient methods and unnecessary tasks. The facts are then organized in an easy-to-understand fashion.
- Review the facts; compare those facts with other like situations; then determine the possible solutions.
- Implement the improved method or practices. This is the trial-adoption phase.

WORK ANALYSIS TOOLS: THE FACT FINDERS

The following are a few of the many fact-finding and solution-getting methods that have become generally accepted by the forest products industry. The cost-reduction potential for each can be substantial.

Work Measurement Techniques

Work measurement and *time study* are broader terms that generally describe work analysis techniques. The following are descriptions of some of the techniques.

Elemental Time Study. A technique developed by Frederick W. Taylor, the elemental time study was originally used to establish should-be work rates for detailed, repetitive jobs. The technique is best suited for evaluating people-paced operations.

In this approach each job or sequence of activities is broken down into elements. Individual elements are timed separately as part of the whole. Each element in the sequence is examined in detail to determine the should-be methods and rates. Figure 19.1 is an example summary sheet for three rough green lumber stackers in a West Coast dimension lumber operation.

The activities of each were broken down into three elements, which were identified as the sum of the total productive activity. The times for each element were individually determined by a continuous observer using a stopwatch. The results were then worked up into normal times for each element.

The normal time per occurrence was then calculated. The analyst subsequently included an allowance for necessary production interruptions (PI). This allowance is only assigned when the cause and the time delay have been clearly identified and the need established.

Also included is a personal time and fatigue factor (P&F) of 7.5%. This factor will be established for each people-paced operation (such as manually stickering lumber courses) and may vary with each operation. (The P&F allowance is seldom used in an automated operation.) In addition, some time studies, although not this one, are leveled or rated to identify level of effort.

While elemental time study has been extensively used in the past, the technique is being employed less frequently for the following reasons: (1) People-paced operations are rapidly giving way to machine control as each workplace is increasingly being automated and (2) the technique is time-consuming and requires judgmental rating and leveling. Elemental time study is being replaced by other methods that represent a more efficient use of study time. These methods include predetermined motion time systems (PDT), work sampling, and group timing technique (GTT).

Predetermined Motion Time Systems. A number of predetermined motion time (PDT) systems have been developed. Each identifies should-be times for manual motions such as reach, grasp and position. The analyst determines the method for an operation, separates the method into elements, applies a predetermined time to each element or portion of an element, and then calculates the standard time as the sum of the elements.

The most common PDT systems include method-time measurement (MTM), basic motion time (BMT) study, motion time analysis (MTA), and universal maintenance standards (UMS). MTM continues to be the most widely used.

Work Sampling. In early 1930 L. H. C. Tippett developed a technique for work measurement that he used to identify production problems in the English textile industry. Tippett subsequently coined the term *Snap Ready.* The technique later acquired various descriptive titles, among them *ratio delay, action delay*, and *work sampling.*

The work sampling technique measures work and non-work time. It focuses in on specific delays and determines the ratio of those delays to the total available time. Figure 19.2 is a summary of one such study. It summarizes edger optimizer

Location Eugene	Mill and operation Lumber	Std. no. 1 of 9
Effective date 3-8-68	Previous effective date 10-1-65	
Machine or job operation Rough green lumber stackers (3)		
Production unit Each course of lumber stacked		

Ele. no.	Element of work description	Normal minutes	Frequency of occurrence	Normal min. per unit
	All 2"–4" Thick lumber —	All lengths		
1	Move new load of lumber onto tilt-hoist	0.645	4/32	0.081
2	Manually break down lumber and feed to stacker table	0.320	32/32	0.320
3	Run elevator down, roll finished kiln car out, and bring elevator up	2.247	1/32	0.070
		TOTAL ⟶		0.471
	PI = 9.2 % – – – – –	– – – –	– – –	0.043

Total time studied normal minutes per production unit	0.514
Standard allowance (if required) 7.5 % P & F	0.038
Total standard normal minutes per production unit	0.552 *
Standard hours per production unit (* × 60)	
BASED ON 32 COURSES / CAR	
Production unit rate per: minute 1.81 hour 109 and shift	

Figure 19.1. Example of time study and its application in standards development.

Downtime causes	Frequency	Minutes lost	% study time	% downtime
No wood in unscrambler	19	8.3	3.2	15.3
Boards crossed on unscrambler	20	7.4	2.9	13.6
Boards would not feed	26	5.9	2.3	10.8
Single board not even ending correctly	25	4.9	1.9	9.0
Edging jammed in picker	5	4.5	1.7	8.3
Outfeed deck jammed	7	4.1	1.6	7.5
Remove debris upstream of edger	16	3.9	1.5	7.2
Two boards at even ending rolls	19	3.8	1.5	7.0
Chain off sprocket of lug chain	1	2.7	1.0	5.0
Board(s) skewed in scan zone	11	2.5	1.0	4.6
Board(s) skewed in infeed zone	11	2.4	.9	4.4
Reposition limit switch	1	2.3	.9	4.2
Board(s) skewed in wait zone	6	1.3	.5	2.4
Misc.	2	.4	.2	.7
Total	**169**	**54.4**	**21.0**	**100.0**

Figure 19.2. Edger optimizer production-time study: work sampling method. (Courtesy Jim Jaderholm)

downtime in a Washington State mill and breaks out the 21% downtime into a number of causes. The first four causes account for nearly half (48.7%) of the total downtime; these same causes also account for 90 of the 169 downtime occurrences. With this information the operator can then focus on resolving the vital few of the many while spending less effort on the other causes.

The total study time consisted of eight separate 30-minute time periods on the day shift. Each time period was selected at random over a five-day work week. During each study period, the observer tallied whether the operation was running and if not, the cause of the downtime.

Each half-minute the observer recorded the activity at that instant; the resulting observations were later tallied. The half-minute interval was selected as one of the sampling parameters because it is convenient to use and speeds data collection.

The study design and sample plan will be simple, complicated or somewhere in between. The repeatability of the product or process will determine the degree of sophistication. The following are steps to accomplish a work sampling study:

Step 1: Define the problem and the information required.

Step 2: Develop the sampling plan. The length of each observation, the timing of each and the time increment to use on the stopwatch can be determined by judgment; usually a trained observer will use professional reference guides to set the plan.

Step 3: Prepare a data sheet on which to record the observations; the sheet should be tailored to fit the task.

Step 4: Walk through the study and debug the format using a trial sample.

Step 5: Collect the data, analyze the resulting information and prepare the results in a format similar to Figure 19.2.

Group Timing Technique. The group timing technique (GTT) combines the best features of both the elemental time study and the work sampling study. It determines both work and non-work times; it also details the actual segments or elements of each.

The observer uses selected time intervals, as in the work sampling study, rather than continuous time study. The work element or non-work activity is tallied at each time interval, such as a half-minute time segment.

A GTT study requires less time than the methods discussed earlier. One observer can time up to 10 or 12 operators working within a reasonable span of observation. The time intervals may be staggered to accommodate the large number of sample observations.

GTT study results can be worked up more quickly than those from a conventional study, and an unskilled observer can turn in study results comparable in accuracy to that of a time study professional.

Work Simplification Techniques

Work simplification procedures combine, simplify, rearrange or eliminate functions into a more efficient sequence of activities. Two techniques or methods are used most frequently. The first is the *flow process chart* and the second is the *multiple activity chart.* Each is easy to use and is frequently introduced to employee participants in better-method or involvement-type participative improvement programs.

Flow Process Chart. A graphic tool, the flow process chart classifies activities by function: operation, transportation, inspection, delay, or storage. The time required for each activity and the distance moved are also documented.

There are two distinct types of flow process charts, the material type and the man type. The first presents the process in terms of events; the second identifies the activities of the individual employee. Figure 19.3, a forklift servicing a hot press and sawline, is an event type.

The chart shown in Figure 19.3 illustrates the sequence of activities and the available idle time. The analyst can use the graphic representation of the work flow both to redesign the job into a more efficient operation and, by showing the proposed work flow on the same format, as a training guide to better illustrate the intended changes. A second, more detailed work simplification technique is the multiple activity chart.

Multiple Activity Chart. A multiple activity chart is a graphic representation of activities performed simultaneously by two or more employees, two or more machines or any combination of men and machines. Figure 19.4 is a man/machine type chart.

A man/machine sequence of activities is shown before and after improvements were made at the infeed of a sawmill edger. This figure graphically shows the before, designated "present method," and after, designated "improved method," sequence of activities and the average elapsed time of each for the edger feeder (man) and the edger (machine).

The original method used the traditional sequence of manual activities; the improved method utilizes computerized hardware to pre-position and feed each sideboard. The improved method reduced process time for each piece from 0.17 to 0.10

Figure 19.3. Forklift serving a hot press, sawline and patchline—sequence of events and available idle times. (Courtesy Jim Jaderholm)

Flow process chart

Summary

	Present		Proposed		Difference	
	No.	Time	No.	Time	No.	Time
○ Operations						
⟡ Transportations						
□ Inspections						
▷ Delays						
▽ Storages						
Distance traveled		ft.		ft.		ft.

No. _____
Page __1__ of __2__
Job FORKLIFT SERVICING HOT PRESS, SAWLINE, PATCHLING
☒ Man or ☒ Material COMBINED
Chart begins PICK UP LOAD AT HOT PRESS
Chart ends RETURN TO HOT PRESS
Charted by J. SADERHOLM   Date 4-22-83

Possibilities — Change

	Details of (✓present/proposed) method	Operation	Transp.	Inspect.	Delay	Storage	Distance in feet	Quantity	Eliminate	Time	Combine	Seque.	Place	Person	Improve	Notes
1	PICK UP LOAD AT HOT PRESS.	●	⟡	□	▷	▽		1	5							
2	BACK OUT AND TRAVEL TO LOAD TURNER.	○	◆	□	▷	▽	30		30							
3	PLACE LOAD IN TURNER.	●	⟡	□	▷	▽		1	6							
4	BACK AROUND TO SWITCH.	○	◆	□	▷	▽	10		10							SAFETY HAZARD! OPERATOR MUST LEAN OUT OF LIFT.
5	PRESS BUTTON TO ACTIVATE TURNER.	●	⟡	□	▷	▽			5							
6	DRIVE FORWARD.	○	◆	□	▷	▽	10		10							
7	PICK UP LOAD.	●	⟡	□	▷	▽		1	3							
8	BACK UP TO BUTTON.	○	◆	□	▷	▽	10		10							
9	RESET BUTTON.	●	⟡	□	▷	▽			3							
10	BACK OUT WITH LOAD.	○	◆	□	▷	▽	10	1	10							
11	GET OFF, PLACE 2×4 BUNKS, GET ON.	●	⟡	□	▷	▽		2	11							WHEN LOADS ARE NOT TURNED, PRESS OPERATORS COULD PLACE BUNKS!
12	TRAVEL TO HOT INVENTORY WITH LOAD.	○	◆	□	▷	▽	30	1	30							
13	SET UNSAWN LOAD IN INVENTORY.	○	⟡	□	▷	▼		1	5							
14	BACK OUT AND TRAVEL TO INVENTORY.	○	◆	□	▷	▽	20		20							
15	PICK UP A LOAD.	●	⟡	□	▷	▽		1	5							
16	BACK OUT WITH LOAD.	○	◆	□	▷	▽	10	1	10							
17	GET OFF, TAKE OFF 2×4 BUNKS, GET ON.	●	⟡	□	▷	▽		2	11							
18	TRAVEL TO SAWLINE WITH LOAD.	○	◆	□	▷	▽	40	1	40							
19	PLACE LOAD ON SAWLINE I/F DECK.	●	⟡	□	▷	▽		1	5							
20	BACK OUT AND TRAVEL TO SAWLINE O/F DECK.	○	◆	□	▷	▽	25		25							
21	PICK UP SAWN LOAD.	●	⟡	□	▷	▽		1	5							
22	BACK OUT AND TRAVEL TO INVENTORY.	○	◆	□	▷	▽	50		50							
23	SET SAWN LOAD IN INVENTORY.	○	⟡	□	▷	▼		1	5							

continued on next page

Flow process chart

Summary

	Present		Proposed		Difference	
	No.	Time	No.	Time	No.	Time
◯ Operations	15	9½				
⇨ Transportations	18	260				
☐ Inspections	0	0				
▷ Delays	0	0				
▽ Storages	3	20				
Distance traveled	665 ft.		ft.		ft.	

Job **FORKLIFT SERVICING HOT PRESS, SAWLINE, PATCHLINE**
☒ Man or ☒ Material **COMBINED**
Chart begins **PICK UP LOAD AT HOT PRESS**
Chart ends **RETURN TO HOT PRESS**
Charted by **J. SODERHOLM** Date **4-22-83**

#	Details of (✓present / proposed) method	Operation	Transp.	Inspect.	Delay	Storage	Distance in feet	Quantity	Time	Eliminate	Combine	Seque.	Place	Person	Improve	Notes
1	BACK OUT AND TRAVEL TO PL INVENTORY.	◯	●	☐	▷	▽	20	20								
2	PICK UP LOAD TO BE PATCHED.	●	◊	☐	▷	▽		1	5							
3	BACK OUT WITH LOAD.	◯	●	☐	▷	▽	10	1	10							
4	GET OFF, TAKE OFF 2X4 BUNKS, GET ON.	●	◊	☐	▷	▽		2	11							
5	TRAVEL TO PATCHLINE INFEED.	◯	●	☐	▷	▽	50		50							
6	PLACE LOAD ON PATCHLINE I/F DECK.	●	◊	☐	▷	▽		1	5							
7	BACK OUT AND TRAVEL TO PL O/F.	◯	●	☐	▷	▽	25		25							
8	PICK UP PATCHED LOAD.	●	◊	☐	▷	▽		1	5							
9	BACK OUT WITH LOAD.	◯	●	☐	▷	▽	10	1	10							
10	GET OFF, PLACE 2X4 BUNKS, GET ON.	●	◊	☐	▷	▽		2	11							
11	TRAVEL TO SANDER INVENTORY.	◯	●	☐	▷	▽	100		100							
12	SET PATCHED LOAD IN INVENTORY.	◯	◊	☐	▷	▼		1	10							
13	BACK OUT AND TRAVEL EMPTY TO PRESSES.	◯	●	☐	▷	▽	200		200							
14	_____	◯	◊	☐	▷	▽										
15	_____	◯	◊	☐	▷	▽										
16	_____	◯	◊	☐	▷	▽										
17	_____	◯	◊	☐	▷	▽										
18	_____	◯	◊	☐	▷	▽										
19	_____	◯	◊	☐	▷	▽										
20	_____	◯	◊	☐	▷	▽										
21	_____	◯	◊	☐	▷	▽										
22	_____	◯	◊	☐	▷	▽										
23	_____	◯	◊	☐	▷	▽										

Figure 19.3. *Continued*

minute. Rather than 5.9 pieces per minute, a bottleneck, the edger capacity was increased to 10 pieces over the same time period.

The multiple activity chart is used both to improve methods in the mill and to determine efficient engineered design for new automated systems and processes. A variation was used by Coe Mfg. Co. to design the charging sequence for small peeler logs into a green veneer lathe using a Model 765 charger and a Model 249 lathe. The multiple activity chart and its many variations are effective tools for both the manufacturing manager and the design engineer.

File

Project No. 4 EDGER MODERNIZATION

Date 2-5-82

Title: MULTIPLE ACTIVITY CHART

Page 1 of 1

PRESENT METHOD

Total min.: .01 .02 .03 .04 .05 .06 .07 .08 .09 .10 .11 .12 .13 .14 .15 .16 .17

Operator		Machine	
ADVANCE PIECE TO EDGER	.03	IDLE	.12
POSITION TO EDGE	.09		
IDLE	.05	EDGER PROCESSES PIECE	.05

% idle: .05/.17 = 29% % idle: .12/.17 = 71%

IMPROVED METHOD

Total min.: .01 .02 .03 .04 .05 .06 .07 .08 .09 .10 .11 .12 .13 .14 .15

Operator		Machine	
PRE-POSITION NEXT PIECE	.05	EDGER PROCESSES PIECE	.05
SEQUENCE PIECES INTO PRE-POSITIONER AND EDGER SIMULTANEOUSLY	.05	IDLE	.05
PRE-POSITION NEXT PIECE	.05	EDGER PROCESSES PIECE	.05

% idle: 0/.15 = 0% % idle: .05/.15 = 33%

Savings in idle time: 29% 38%

Improvement in pieces/minute 7.5

Figure 19.4. Multiple activity chart comparing present and proposed method. (Courtesy Jim Jaderholm)

REQUIREMENTS FOR USE OF WORK ANALYSIS
TECHNIQUES

Each of the methods and techniques described are cost-saving devices if used effectively. The following are requirements that will ensure favorable results:

- Clearly define and document the process or job description to provide a reference point for later comparison.
- Establish and detail the required or expected quality level to promote better understanding.
- Establish reference points within the workplace layout. For example, establish the operator's location in relation to machine, materials and tool placement.
- Work out machine setup specifications, speeds and machine functions both before and after the analysis.

In addition, the job or process redesign should consider the needs of the individual employee. Efficiency is important; but it's also frequently possible to increase efficiency by minimizing or eliminating monotonous, hectic and mentally strenuous activities.

Higher work satisfaction and participative contribution usually result in greater motivation and therefore improved efficiency. The end result is more cost dollars that fall past each potential stopping place and then descend to the bottom line.

REFERENCES

Ager, B. 1975. Improvement of Mill Working Conditions—Objectives, Benefits and Costs. *Nordic and North American Sawmill Techniques*, Ch. 15, p. 184. Proceedings from the Second International Sawmill Seminar. San Francisco: Miller Freeman Publications.

Chase, R. B., and Aquilano, N. J. 1981. *Production and Operations Management: A Life Cycle Approach*. 3d ed. Homewood, Ill.: Richard D. Irwin.

Heyel, C. 1979. *The VNR Concise Guide to Industrial Management*. New York: Van Nostrand Reinhold Co.

Holemo, F. J., and Dyson, P. J. 1971. Ratio-Delay—A Method for Analyzing Downtime in Sawmills. Abstract of speech presented at Session 16, "Production Management," of the 25th Annual Meeting of the Forest Products Research Society, June 30, 1971, Pittsburgh, Pa. *Forest Products Journal* 22(8):56 (August).

Maynard, J. B. 1971. *Industrial Engineering Handbook*. 3d ed. New York: McGraw-Hill.

United States Army, Management Engineering Training Agency. 1981. *Defense Management—Joint Course*. Vols. 1 and 2. Rev. ed. Rock Island, Ill.

White, M. S. 1980. *Procedures for Analyzing Sawmill Performance*. Sandston, Va.: Lumber Manufacturers' Assn. of Virginia.

Twenty
Energy Management

The 1974 energy crisis was a traumatic impetus for change in the forest products industry. Forced curtailments and soaring energy costs combined with a lackluster economy led one manager to observe, "You just can't afford to make plywood anymore using fuel oil for the boiler and natural gas for the veneer dryers."

Fuel oil prices have tripled and quadrupled within recent years; natural gas supply contracts, many negotiated when natural gas was abundant, have been renegotiated with price increase multiples of eight to ten. Electricity costs have followed the upward trend. And the end is nowhere in sight.

The manager's task is to define the continuing energy cost reduction task and then continue to offset rising costs through wise and selective use. The experience of an industry trade group is an example of what can be accomplished through group and individual effort.

The Wood Products Industry Energy Conservation Program, initiated by the Council of Forest Industries (COFI) of British Columbia, reported a 15.7% reduction in electrical consumption for its member lumber manufacturers during the period between 1978 and 1982. Natural gas consumption was reduced 32.9% during the same time period.

Overall the 65 mills included in the program, or 30 companies representing roughly half of British Columbia's lumber production, were able to effect sizable gains in each individual unit. These same gains can be achieved whenever an organized approach to energy conservation combined with the substitution of lower cost energy alternatives is undertaken by the individual company or mill. Knowing where and how to achieve gains are prime energy cost control considerations.

WHERE TO LOOK FOR SAVINGS: ENERGY
SYSTEMS AND SUBSYSTEMS

Opportunities to reduce energy costs are many and varied. A detailed and thorough approach will first identify the systems and subsystems; opportunities will be classified within this framework.

A hog fuel boiler is usually less costly than other steam-producing alternatives.

Primary energy systems include the steam or heating system, the compressed air system and the electrical systems. Each primary energy system can be divided into three subsystem classifications: source, distribution and end-use. The boiler is a visible example of a source subsystem.

Boilers

Air-to-fuel ratio, moisture content of the fuel, size and character of the incoming fuel, and fuel feed rate are primary factors that affect a boiler's combustion efficiency. The following are ideas to increase efficiency and shave costs.

Air-to-Fuel Ratio. An insufficient proportion of air results in incomplete fuel combustion and high pollutant emissions. A high proportion of air in the air/fuel mixture results in heated air exiting the stack in large quantities, thereby wasting heat and fuel. A tight, clean boiler with unrestricted air flow will burn cleanly and efficiently if operated with the just-right amount of air and fuel.

Improved results can be effected. Periodic cleanup, inspection and repair are musts. Other aids include accurate instrumentation and installation of an oxygen analyzer to regulate draft and fuel intake so that each is optimum.

Fuel Moisture Content. The BTU value of wood residue is a direct function of its moisture content. Heat energy is expended in drying the fuel prior to burning. As a rule of thumb, the moisture content of hog fuel cannot exceed 50% and result in efficient combustion.

Storing fuel in a covered storage area is one help; fuel dryers or presses are occasionally used to obtain the just-right moisture content.

Size and Character of the Fuel. The time it takes to achieve combustion will vary with the size and character of the incoming fuel. Reducing the size and obtaining a homogeneous mixture will shorten burning time and ensure more complete combustion. Shredders and hogs are customarily used to size the rough hog fuel that develops from bark and wastewood chunks.

Fuel Feed Rate. Wood fuel should be fed at the just-right rate; too low a rate will not maintain the combustion process, and too rapid a rate will allow unburned wood to pile up in the combustion chamber.

The tips cited above will result in combustion efficiency improvements and reductions in fuel use of 5%, 10% and more. Just as important is the consistent and even steam flow that the boiler will produce.

Air Compressors

An air compressor is another example of a source subsystem; its air lines and valves are part of a distribution subsystem. All compressed air systems, particularly those operating at high pressures, will have leaks that can waste as much as 20% of the compressed air they generate. Table 20.1 shows the cost impact of small leaks. Leaks occur at valve packings, quick disconnects, regulator handles, blow-off nozzles, pipe connections and around areas of vibration or wear from adjacent components.

Compressed air leaks can best be detected during weekends or shutdown periods such as lunch when the noise level overall is subdued or absent. Most of the

Table 20.1. Compressed Air Leakage Costs

Hole size (in.)	Cfm at 100 psi	Cost per yr
1/32	1.11	$ 55
1/16	4.2	210
1/8	16.8	840
1/4	67.5	3,375
1/2	270.0	$13,500

Note: These costs assume 24-hour daily operation of the compressor and electricity costs of $0.05 per KWH.

discovered leaks can be repaired with normal maintenance procedures. Constant follow-up and repair are the key to minimizing energy losses in the compressed air transmission lines as well as in other transmission systems, such as the steamline counterparts.

Lighting

Lighting hardware is an end-use subsystem. Replacing existing fluorescent lamps with lower wattage energy-efficient types is an energy conservation alternative. Generally, since the new lamps are of lower light output, the light level will be reduced by about 5% to 11%. However, the wattage reduction will range from about 15% to 20%. Figure 20.1 outlines suggested applications for various types of industrial lighting.

The actual efficiency of each type is measured in lumens per watt. A lumen is a measurement of the light developed; a watt is a measurement of the energy used to generate that light. The higher the number, the more efficient the industrial lighting type, and the more light that is obtained for the energy consumed.

Table 20.2 identifies the relative efficiency of light hardware types. Fluorescent lighting is rated at 75 to 85, with the sodium systems ranging upward to 183. The table also illustrates the comparative costs for different light sources and indicates the wide disparity in operating costs among the various lighting types. Introduction of the newer light types can sharply reduce the lighting bill for the plant and adjacent work areas. Lighting hardware is an end-use subsystem and is a prime source for cost reduction.

ORGANIZE FOR COST REDUCTION

The manager will avoid the random purchase of "newfangled gadgets" in a hit-or-miss attempt to reduce energy costs. Instead he will check the major sources of energy waste, some of which have been described, and focus on reducing the costs generated by improper and/or excessive operation, poor maintenance of machinery and use of obsolete or costly equipment.

Conservation of energy involves more than effective engineering and frugal purchasing practices; it also involves changing habits and practices. The following is a management technique that will both effect initial savings and change habits.

Industrial lighting type	Applications
Incandescent	• Tasks requiring small sources • General-purpose task lights • Small storage areas • Special applications requiring low-cost, warm-color lighting
Fluorescent	• General-purpose task lights • All low-ceiling work areas • High-ceiling areas where low initial cost is a prime consideration • Limited outdoor use for temperatures above 60°F
Metal Halide	• All medium to high-ceiling work areas where good color rendition is required • Outdoor floodlighting and security lighting
High-Pressure Sodium (orange color)	• High-ceiling areas and exterior use where color rendition is not important and *high* energy efficiency is required • Warehouses, parking lots, security lighting, loading docks, shop and assembly areas
Low-Pressure Sodium (yellow color)	• High-ceiling areas and exterior use where color rendition is not important and *very high* energy efficiency is required.

Figure 20.1. Industrial lighting applications.

Table 20.2. Relative Lighting Efficiency and Comparative Costs for Various Industrial Light Sources

	Lumen per watt efficiency		Average lamp[1] life	$ per 10M lumens (light) output for 1M hrs at power cost per KWH of:		
	Range	Avg	(hr)	1¢	3¢	6¢
Incandescent	17–24[2]	20	1,000	5.00	15.00	30.00
Fluorescent	75–85	80	12,000	1.33	4.00	8.00
Metal halide	80–115	100	10,000	1.00	3.00	6.00
High-pressure sodium (orange color)	70–140	110	15,000	0.91	2.73	5.45
Low-pressure sodium (yellow color)	130–183	155	18,000	0.65	1.95	3.87

[1]Large variation depending on size, number of starts, and application.
[2]Also quartz.

This technique is twofold; it initiates an energy audit by professional technicians while utilizing and involving operating personnel such as foremen, maintenance employees and machine operators. Maintenance employees and foremen, for example, usually work throughout the mill or department. The scope of their activities puts them in a position to identify and correct energy losses.

Energy Program Task Group

One step in effecting an energy cost reduction program is to organize a task group of in-house personnel; then appoint an energy coordinator to provide leadership. This energy coordinator functions both as the discussion leader and as a program tracker.

A task group of six to eight employees is best; ten is the outside limit. The individuals selected should represent a mix of skills, including mechanical and electrical. Figure 20.2 is a schematic that highlights the action steps, steps that focus on low-cost solutions rather than "big bang" capital projects.

The task group meets periodically to discuss and explore specific energy uses. At each meeting the energy coordinator cites a specific application and directs the discussion to particular operations. Notes are taken and specific assignments (who does what by when with what results at what cost) are developed. A minimum of 15 minutes is suggested for each topic; the discussion can continue until the situa-

Figure 20.2. Energy program task group: action steps.

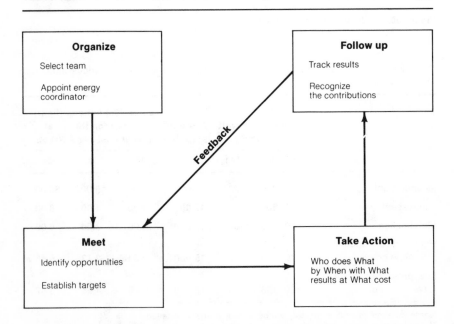

tion is fully explored. Occasionally a topic may be deferred until additional facts are obtained.

The implemented ideas can range from simple solutions (such as shutting off a lily pad chipper when it's not in use) to more technical solutions (such as adding additional capacitors to the electrical system).

The Energy Audit

The second part of an energy cost reduction program is to initiate an energy audit by professional technicians. This audit should focus on the three major energy systems: steam, air and electrical.

The first step is an energy survey of each system. For example, the electrical audit will document both the amount of electricity used (energy charge) and the kilowatt (KW) demand charge. The latter identifies the maximum amount of electricity used (in KW) over a specified period of time stated in the rate. Other special charges will also be identified.

Next, a connected horsepower study will be prepared and correlated with the amount and demand for electricity. This reconciliation will take into account other electrical consumption, such as heat/air conditioning and lighting.

The resulting facts are then studied and translated into cost-cutting opportunities. This may require the assistance of a public agency such as an energy office or extension service; it may also involve the assistance of the utility or fuel company, outside consultants such as HVAC engineers or the suppliers and manufacturers of mill equipment. For example, a motor user's question may require a detailed answer of the sort available from General Electric Co.'s new Energy Saver Computer Program.

The answers are then organized into a detailed program. Goals are established, with the priority given to no-cost and low-cost changes. A minimum acceptable return on investment or some other financial yardstick is used when applicable. Approved projects are then delegated for timely implementation.

The people program (Energy Program Task Group) and the formal energy conservation program may be initiated separately or concurrently. Each is designed to complement the other; one may perform the spadework for the other. The overall program goals are to identify ideas that will reduce energy costs and use and to gain commitment to changing habits.

The end result is a mill that operates more smoothly with lower costs and with involved employees—necessary ingredients to continued operation.

REFERENCES

American Plywood Assn. 1980. *Trimming Your Energy Costs in the Plywood and Veneer Industry*. Research report. Tacoma, Wash.

Hahn, T. M., Jr. 1980. The Role of Wood in Our Energy Future. Presentation at Bio-Energy '80, Atlanta, Ga., April 21, 1980. Portland, Oreg.: Georgia-Pacific Corp.

Horovitz, B. 1981. Cutting Your Energy Costs. *Industry Week*, February 23.

Lambert, D. M., and Stock, J. R. 1979. Organizing and Implementing the Corporate Energy Plan. *Michigan State University Business Topics*, vol. 27, no. 3 (Summer).

Price, S. G. 1973. *A Guide to Monitoring and Controlling Utility Costs*. Washington, D.C.: BNA Books.

Westergaard, B. 1982. The Evolving Sawmill. *Logging & Sawmilling Journal*, September:32.

Section Six
MANAGING CHANGE

Question, question and question your business strategy and practices; search for new ideas that can expand or improve the product; check manufacturing practices to ensure that each is being followed consistently and that each is the just-right one for the mill.

CHAPTER TWENTY-ONE

Twenty-one
Operating the Mill in a Lean Market— Or How to Avoid the Cut

Professional football fans are familiar with the procedure that is used to fit the size of the preseason squad to the required league team roster; it's called the cut! This procedure releases extra players from the squad shortly before the season begins. Unfortunately, a forest products operator in a tough market is in a similar position to a professional football player.

Traditionally during down cycles, the number of operating mills is pruned to reflect real demand; the economic climate also tests the ability of the remaining operators to exist. The following are a number of "old hat" ideas that make it more difficult to prune your operation. They include tips on operating strategy, communication and spending. Figure 21.1 summarizes each of the tips.

USING THE PLAYBOOK

Establish credibility by achieving stated goals in production, cost control, reporting requirements, forecasts, plans and other related management tasks. Do what you say you will do. In no event allow any surprises. Safety, environmental, maintenance, and other important ongoing programs should be audited closely so that these support programs don't demand unwelcome attention that may distract from the ongoing business. In addition, focus on quality.

It's a buyer's market. Product defects that would normally go unnoticed become highly visible as the warehouse personnel, the jobber and even the carpenter get picky. It is not uncommon to have a load of product rejected by the customer because of perceived quality problems such as crook in lumber and core voids in plywood. In one case a would-be buyer concluded, "I can probably buy it now cheaper anyway."

That buyer's statement is an extreme example, but little things such as stencil appearance and band placement all become highly visible to the user and potential customer.

Question, question, and question your business strategy and practices. Do not assume anything. Search for new ideas or projects that can expand or improve the product or product mix. Then check all manufacturing practices to ensure that each is being followed consistently and that each is the one that is just right for the mill.

Significant changes in production strategy should be tested using simulation or optimization techniques (see Chapter 6) prior to implementation in the mill. Avoid seat-of-the-pants decisions that can seriously damage short-term profitability in a good market and sink the mill into red ink in the lean years. And don't try something new just for the sake of change.

When asked how he survived and prospered in the lean years, a Nevada cattle rancher conceded, "We do things the old style, but the reason we do it is it's the most efficient, cost-effective way of doing things" (*Wall Street Journal*, 1982). And the forest products business is not much different. We continue to do many things the old way because it is the best and the most cost effective.

COMMUNICATING THE PLAY

The plant grapevine is a marvelous communication tool . . . as long as it operates for the manager; and facts are the best tools to help the grapevine work for you. One-on-one communication each day from the "same sheet of music" by the immediate supervisor will improve morale and increase confidence in local management. A weekly staff meeting with the manager and his key personnel to review the month-to-date facts is a good vehicle to ensure the best decisions by each member of the management team. No less important is timely communication with support groups such as sales and distribution.

Work closely on at least a daily basis with mill sales representatives to ensure that communication takes place and that all orders are given a thorough review for suitability. And keep your order options open.

A mill that has multiple play options stays on line longer than a mill that is dedicated to one or two items, such as underlayment or rough sheathing for the plywood producer and dimension lumber for the sawmill. Grocery store type orders

Figure 21.1. Tips for avoiding the cut.

• Using the Playbook	Find the just-right operating strategy and implement it.
• Communicating the Play	Communicate the strategy and operating plans to each participant.
• Coaching the Play	Bird-dog the manufacturing process.
• Manning the Roster	Maintain and increase productivity with fewer man-hours.
• The Third-Down Conversion	Control costs. Examine each expenditure.
• Avoiding the Blowout	Avoid capital investments that are not cost effective.
• So You Didn't Make the Playoffs	Make the best of a shutdown or curtailment.

with lots of 1-in. lumber or ¼-in. panels may be a more profitable option than a four-day week or no week at all.

Don't make items you can't sell. Seconds, culls, No. 4 lumber and shop panels are the most visible examples. A serious effort should be made to determine the real cause of falldown or degrade and then to avoid manufacturing it.

Late orders are a "no-no"! One mill sitting on late orders while another within the same company is curtailed for lack of orders spells additional dollar losses for the company. The pricing policy is well established in most companies, and even fixed-price orders are subject to renegotiation if they are late.

COACHING THE PLAY

The manager who takes care of the small things will find that the big things will take care of themselves. Paying attention to details is even more important in the hard times. This approach will have two results:

- A consistently uniform product will be produced with the least material, cost and labor ingredients.
- When problems occur—and they will—the manager/supervisor will catch them when they are small rather than after they have grown into bigger, less manageable problems.

The "take care of the small things" principle does not imply that the manager deals with details rather than problems. In addition, it should not be interpreted as an "ego trip," a chance for the manager or supervisor to demonstrate his expertise for his own satisfaction. The intent of the principle is for the manager to use operating controls and applied experience to seek out big problems in the formative stage. A West Coast mill is an example.

During the early 1960s, a veteran superintendent was guiding a veneer plant through the throes of start-up. He had a number of less experienced supervisors on each of the night, swing and graveyard shifts. The graveyard shift supervisor and his crew were especially young and aggressive. Lathe and green clipper production on that shift was considerably higher than on the other two shifts, particularly than on the day shift, which was supervised by the superintendent.

One morning the graveyard supervisor mentioned that fact once too often. "Pete," commented the superintendent, "you are saving quarters in unit costs and giving away dollars with the small things you are missing."

The next two hours were spent examining the small things still in evidence from the prior shift. Rough peels, poorly stacked loads, inadequate housekeeping and incorrectly placed spacer sticks were a few of the many small things. The combination of small things added up to big things as the downstream costs of degrade, broken full sheets, downtime and lower-than-should-be production rates were all considered.

Most mills have more than their share of the small things, small things that are lived with in the good times for no apparent reason other than neglect. A market trough provides more than enough incentive to responsible personnel to get each corrected.

MANNING THE ROSTER

Staffing and man-hour control are key functions of management. One manager of a large sawmill, for example, began reviewing the daily man-hours and crewing for each department. Not surprisingly, the following week witnessed a sharp drop in man-hours for the same volume and product mix. This is a good example of managing by emphasis, that is, ask the questions where the big dollars are. And be aware of man-hour efficiencies, particularly when the order file is lean.

Keep the volume per man-hour up, but do it on as few operating hours as possible. There is a tendency to stretch a slim order file to fit the hours available. Unfortunately that is a good way to demonstrate to the bank or the home office that the mill is unprofitable even with ample orders and good prices. Using the lean times to break existing performance records is a good way to maintain employee morale. It is also an excellent way to maintain top management support.

Man-hour control is a key management function.

THE THIRD-DOWN CONVERSION

Control of expenditures is something like a third-down conversion—you either make it or you don't. And there is a lot riding on each spending decision.

Eliminate the wants and needs right off. Only the musts should get purchased, and then only in the just-right quantities. Keep company dollars in house; make or make do with the materials on hand as much as possible. This may not be convenient but it is profitable.

Make the best use of your resources. Use the available time of hourly personnel, maintenance workers and salaried employees for projects and maintenance that will reduce future costs or increase mill operating efficiency. Rebuild rather than purchase new whenever feasible.

Concentrate on gleaning the best information from yourself and your people. Previously acquired knowledge and skills can be used productively to search out new ideas and less costly, better methods. Formally and informally ask the question at each cost center, "How can this operation be combined, rearranged, and/or simplified; do we really need it?" This includes an analysis of those things that are normally considered fixed costs, such as communication devices, delivery services and other ongoing periodic expenditures. This may also include a review of big-ticket items like energy costs and power contracts.

AVOIDING THE BLOWOUT

"Increase the investment, increase the efficiency and then we will be a low-cost producer ready to compete with anyone" is the theme. Unfortunately, the idea sounds better than it usually is.

A manufacturing facility dealing with a commodity product in a cyclic market can rapidly find itself overcapitalized and forced to continue operations when it should be curtailed or shut down. The trick is to determine and accept a tradeoff between improved variable costs and disproportionately higher fixed costs.

Frequently a producer will purchase expensive yield-increasing and labor-reducing equipment when equivalent improvements are available through improved manufacturing control and better methods. These comments are not intended to discourage worthy capital investments. The wise operator will ascertain that the investment and its associated costs (such as installation, start-up and fixed) will be justified based on expected savings.

The investment must be part of a total operating strategy rather than a substitution for marginal manufacturing practices.

SO YOU DIDN'T MAKE THE PLAYOFFS

Shutdowns and curtailments occur, and the task for the operator is to make the best of a difficult situation. The best includes taking time to better position the operation for the future. It also includes assisting other company operations that are still able to operate.

Maintenance can be intensified in most cases at little actual out-of-pocket cost. Frequently, the maintenance jobs that have been put off or postponed are those tasks that are time-consuming and require extensive downtime or man-hours to complete. Too often rush jobs that occur around a normal production schedule have a cost penalty, a penalty that can be avoided if those jobs are done during curtailment. The key is forward planning.

Implementing concentrated supervisory training programs also requires forward planning. This forward planning identifies valuable programs that need to be completed when time is available. This may include advanced instruction in first aid, safety training, basic supervision and human relations. Many advanced courses are too time-consuming to accomplish during a normal operating schedule. This training may also include specialized instruction in state-of-the-art manufacturing processes and the latest computerized process control technology, which may be installed soon or in the future.

The tips described in this chapter may not result in a mill "missing the cut," but each will give the operator an edge. A mill needs that edge.

REFERENCES

Blackman, T. 1982. Modernizing Helps Mills Survive the "Depression". *Forest Industries* 109(9):34 (August).

Burck, C. G. 1982. Can Detroit Catch Up? *Fortune*, vol. 105, no. 3, (February 8).

Smith, K. 1982. After the Mill Closes. *Old Oregon* 62(2):1 (September). Eugene: University of Oregon.

Wall Street Journal. 1982. Old West: A Nevada Ranch Survives in Lean Times. 69(15):1 (January 22).

Twenty-two
Integrating New Technology into the Established Mill

Introducing new technology into an established mill is something more than installing a new piece of equipment. It should be the end product of an overall strategy, a strategy intended to introduce systems rather than machines.

The equipment design process, heretofore directed by an engineer or master mechanic, has evolved into a multidiscipline task that is increasingly more specialized and more complex. The task of the operating manager is to determine how this high technology can be utilized; and utilized in a fashion that will produce the results intended. Five action steps are required to successfully effect changes in the established mill. Let's examine each in sequence.

ACTION STEPS FOR NEW-TECHNOLOGY PROJECTS

Identify Objectives

Equipment and processes seldom wear out before each is obsolete; therefore, the capital investment program will be something more than equipment replacement. Operators with one, two or just a few plants may focus on competing for logs and markets in order to survive. Their much larger corporate cousins may define a broader yet specific need such as: "To improve earnings 20 percent with an investment of 38 million dollars over the next four years in fifty-five technology projects utilizing five state-of-the-art technologies" (Napier 1982, p. 2).

This statement recognizes the objectives, or goals, of the business or corporation. It is broad enough to allow flexibility when determining the details, yet detailed enough to provide the framework for determining the specifics as the technology evolves. The key questions that each capital investment strategy statement should answer: What is the strategic direction of the business? What are the specific performance goals that need to be achieved? What are the options for achieving each goal?

The answers might be something like this statement: Area log costs are expected to increase 25 percent overall within 30 months. To successfully compete for these logs and increase profitability we will be required to raise the current lumber recovery factor (LRF/BM of lumber per cubic foot of logs) from 7.8 to 9.0 over the same time period. This may require equipment modification and/or replacement at the bucking operation, headrig, edgers, resaws and green trimmers.

The actual statement outlined and structured by the manager or managers will identify overall accountability for the successful implementation and achievement of the project goals. The task is to obtain sufficient information from the line and staff participants to provide the best answer; involving them will also foster a participative vested interest in the project goals.

Participative input will largely overcome the following obstacles to successful implementation, which often appear at critical points during the project cycle:

- A lack of direction, understanding, commitment on the part of participants.
- A failure to plan sufficiently for the projects by limiting input.
- An inadequate recognition of the current operating environment.

Failure to overcome these obstacles is costly. The symptoms are slow, sloppy startups, costly budget overruns, and a generally difficult time in achieving performance objectives. Identifying the objectives, then, is doing something more than providing direction. It is providing direction in such a fashion that the subsequent steps proceed smoothly. The next step is to investigate the various technologies that may have the potential to satisfy the project planning objectives.

Investigate Technologies

Step 2 begins with further definition of the possibilities. For example, a primary breakdown unit in the sawmill may be designated as a leverage point to achieve all or part of the objectives.

The planner will recognize that he must obtain the best information on the options available to replace or upgrade. Additionally, he must correctly identify what he can expect in logs and markets for the present and in the years ahead. This implies more than a casual conversation with marketing and log procurement. It identifies the critical need to cover all bases in planning and to prepare a detailed analysis for each element. The detailed analysis then provides the basis for identifying the technology that will best meet the objectives.

The identification process will begin with a state-of-the-art survey to determine just what is available in the trade or an investigation of a vendor's capability to fill a need. It should be remembered that the results of the investigation reflect the options at a point in time that may pass in the current technology explosion. Commented a mill operator, "I just bought it two years ago and already I am having to replace major components because something better has been developed in the meantime."

A leading electrical equipment vendor reinforced this comment: "The technology is changing so rapidly that they have developed something new before the current generation is in place. We are finding that each new generation is more powerful than the former; in fact, each advances our capability to develop even more powerful generations."

At times this sheer momentum of development, especially in process controls, can overwhelm the investigator. He is inclined to say, "Why not wait until this development process peaks and then I will buy." Unfortunately, that is not likely to happen anytime soon.

The investigator will need answers to the following questions when investigating technologies:

- Will the currently available technology meet the objective? What developments can be expected in the future? Would retrofit be feasible?
- How reliable is the technology in the field? How many units have been installed? How successful has each been? What are the facts? Who is best able to give me these facts (mill visits, phone and in-person interviews with the operator and/or experience of others)?
- How reliable is the vendor? What is his track record for this particular application? What has it been in the past with other new technology? What is his service capability? Will he be able to provide the help required in a reliable and timely fashion? What is his capability for training my personnel in operation and maintenance? What provisions is the vendor prepared to make to keep the equipment up to date as new developments occur?

As is the case with this gang saw, equipment and processes seldom wear out before each is obsolete.

From this bandmill, primary breakdown unit at a South Carolina mill, multiple cuts flow to a 5-ft Salem resaw, and single cuts are sent to the adjacent Salem board edger. (Courtesy *Forest Industries* magazine)

The audit will determine the available options. It also provides the foundation for the next step.

Select the Best Fit

The screening and justification techniques are similar to those used in a gross feasibility analysis (Chapter 9). The task is to evaluate and compare the available options and then select the one that will provide the best fit. Questions to be asked and answered include:

- In what way does the new technology fit into the company or individual business unit? What is the probability for achieving and even exceeding the results expected?
- What is the future for the technology selected? Will it successfully mesh with the current and expected equipment or process configuration? Is the scope of the application wide enough to contribute to the entire system? What will its expected rate of obsolescence be?
- Is the proposed technology the first or the tenth copy? Will there be extraordinary debugging costs? Can these costs be amortized as part of the individual investment or will they have to be amortized as the first of a number of like installations? What is the degree of risk associated with the project?
- How will the project change the statement? At the initial company location? At succeeding locations? How does it fit the strategic goals of the company? What will be the expected return on capital employed (Chapter 9)? What will the payback period be?

The answers to these and other questions will determine the capital costs, the benefits and the payback period, and the risk associated with each technology alternative by vendor. The "just-right" selection will be the end result of these analytical questions.

Plan and Implement

Figure 22.1 illustrates the project planning and implementation process. There are four differentiated activity groupings; the involvement over time is designated by the bar coding of each activity. The solid bar indicates primary activity or responsibility. The broken bar indicates the phasing in and out of activities.

Engineering and Construction. The engineering and construction activity begins with the inquiry and planning stage. It further includes the detailed planning and construction that will result in the completed project. There is generally heavy participation with equipment vendors and with mill personnel. Concentrated activity starts with the technology selection and ends with the online project.

Startup and Debugging. The team approach is increasingly used to introduce new technology. The team participants represent a mix of skills including but not limited to production supervision, skilled or semiskilled employees who have leadership skills and have worked previously or intensively trained with the new technology, and skilled installation personnel. This team concept is usually applied in either of two ways: (1) Turnkey operation, which is then turned over to the operating personnel when startup is completed; (2) An organization that provides assistance and is phased out as the local group assumes greater control of the technology application.

Ongoing Operation and Maintenance. A project has not been successfully implemented until the operating and maintenance personnel demonstrate that the

Figure 22.1. Project implementation.

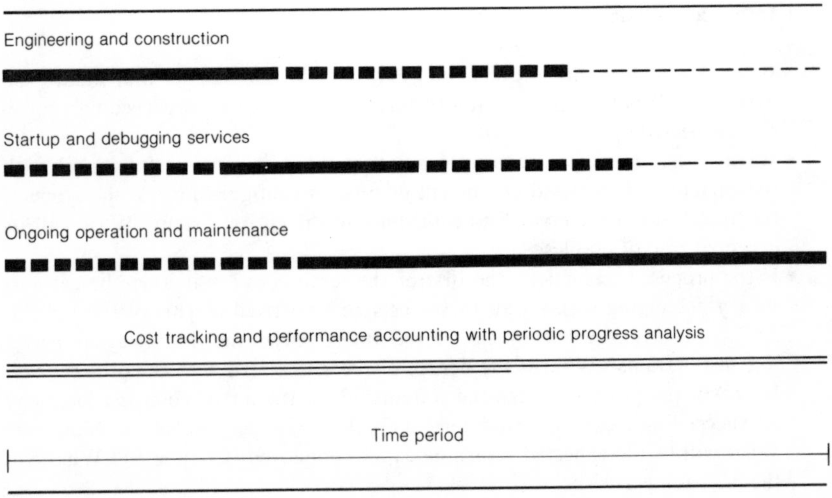

Engineering and construction

Startup and debugging services

Ongoing operation and maintenance

Cost tracking and performance accounting with periodic progress analysis

Time period

project will function effectively and efficiently, as intended. The transition to on-going operation and maintenance begins with the planning phase; the key operating and maintenance personnel need to be a part of the detailed planning. In addition, an intimate knowledge gained while observing construction can result in a smoother startup and quicker response to problem areas as the project is brought on line.

Cost Tracking. A demanding activity, cost tracking records the actual technology installation and the associated startup and debugging costs. It is used as a management tool to determine areas of opportunity that need to be investigated.

Track and Follow Up

A network planning model such as CPM or PERT (see Chapter 6) is used to identify the sequence and timing of activities as the project is fitted together. This model combined with the cost accounting and performance accounting activities will be effective tools for indicating when corrective action is warranted.

The five action steps outlined begin as the objective is identified and end with the tracking and follow-up sequence.

PROJECT CHARACTERISTICS

A successful project will have the following attributes:

- Objectives are clearly defined and are understood and agreed upon by both mill and project personnel.
- Well-defined roles are outlined for each project participant, including mill operating personnel.
- Contractors, when used, have the skill levels required and supply an adequate number of continuing on-site personnel. Shifting of contractor personnel is avoided.
- The equipment vendor is dedicated to a successful installation and has dedicated a sufficient number of skilled personnel for the duration of the installation and subsequent follow-up.
- The software is thoroughly checked with in-mill personnel and electrical drawings are verified before installation.
- An attainable CPM/Gantt-type schedule has been formulated and agreed upon by mill and project personnel. The schedule dedicates project personnel for the project duration, with a coordinated round-the-clock installation schedule when required.
- The completed project is released to the mill with continuing detailed follow-up by the project manager until the project is successfully integrated into the mill operating environment.

The least successful projects are characterized by the following:

- "Homework" done on site
- Shifting of project, contractor and vendor personnel on and off the site

- Lack of information about details of the project on the part of mill and project personnel
- A more complex and complicated installation than required
- A loosely defined installation schedule

The prime objectives are to get the project in place on time and within the budget and to have the new technology perform its intended function. And that's a pretty tall order in an ongoing industrial environment. But the role of the operating manager is to do just that.

REFERENCES

Hall, A. 1962. *A Methodology for Systems Engineering.* Princeton, N.J.: Van Nostrand.

Napier, T. W. 1984. Implementation of Process Control Technology. *Computer Automation for Sawmill Profit*, p. 13. Proceeding 7333, FPRS Conference, Norfolk, Va., October 4–6, 1982. Madison, Wis.: Forest Products Research Society.

Twenty-three
Start-up and
Turnaround Situations

The sun had dropped behind the edge of the lake on a late October afternoon. The foremen's meeting had ended; the participants had moved from the meeting room to an out-of-the-way corner in a northern Idaho restaurant. The refreshments were cold and the talk warm as the veterans and would-be veterans told and listened to stories about other mills in other places.

A 35-year veteran ended the reminiscing and summed up the discussion just prior to leaving: "You know, mills are somewhat like women. . . . Each has a different disposition . . . and no two are just alike. A mill start-up is something like an old man marrying a teenage bride; he just can't take anything for granted and he has to teach her everything she needs to know." The talk continued: "Shaping up a sorry mill is something like what a man has to do when he has been gone from home six months too long. She's probably picked up a lot of bad habits that need broke."

This analogy has a kernel of truth, whether or not the reader agrees with the domestic conclusions. The problems are real and can be avoided during a mill start-up or turnaround.

THE START-UP

The plan is the latchkey to a successful start-up. It should include capital and construction details that are tailored to fit the original justification. The capital request then becomes the source document for establishing the preoperating and start-up plans.

Content of the Plans

The preoperating and start-up plans will include both on-site activities plus off-site needs (such as housing, schools, medical services and other community services). This off-site infrastructure is particularly important to the forest products operator. Well-sited mills are often located in rural areas away from urban development.

The start-up plan should be finalized prior to construction completion.

The on-site start-up plan has several important areas of emphasis. These include market identification, log profile analysis, labor and skill needs, log transportation methods and costs, by-product markets or uses, and equipment types. Others, such as environmental and safety concerns, are given no less attention.

Each element of the start-up plan is prepared in detail and carefully documented. The resulting information forms the statistical base for constructing the zero-based tactical or start-up plan. The start-up plan prepares the operator for implementing and tracking ongoing start-up activities.

Start-up plan preparation is not a one-man show; it is a participative process that draws input from each line and staff activity. This information is evolved into a succession of discussion drafts until the group has arrived at the best-fit plan. The best-fit plan should meet or exceed the performance requirements necessary to achieve the project justification.

The start-up personnel closet themselves with the construction leaders and correlate the plan with the forecasted completion schedule. This correlation is relatively easy to do for a project that has been based on a PERT or CPM schedule and that has maintained timely completion dates for individual project segments.

Staffing Requirements

The emphasis then refocuses on staffing requirements. The supervisors and key machine operators are identified and introduced to the project. Each supervisor is assigned to a specific mill area to ensure that the little things so important to the efficiency of the operation are taken care of as the construction and equipment installation proceeds. Key hourly employees, such as the kiln operator or the lathe operator, work closely with the assigned supervisor. The payoff is a close working relationship and an intimate knowledge of the mill.

The supervisors and key hourly employees take the overall start-up plan and break it down into a detailed machine-by-machine routine. Schedules, standard operating procedures (SOP) and specification manuals are prepared and analyzed closely. The goal is to rehearse step by step the situations that can and will occur in a start-up situation and on into the normal operation. A thoughtful review of the applicable commercial/product standards and the defined policies/procedures of the company are also completed individually and collectively during this period.

Maintenance personnel are selected and folded into the construction crew whenever possible. Some forest products companies will use this time to train the incoming maintenance personnel, offering formal and structured training such as that provided at nearby community colleges or by state-sponsored business development services. Others will farm experienced personnel out to necessary projects that are outside the scope of the contractor's designated activities.

Hourly employees are introduced to the workplace at predetermined intervals prior to the start-up. Formal training should occur both in the classroom and at the jobsite. This formal training will introduce the employee to the company and its policies. It will pre-answer the normal questions that are certain to occur. A number of industry and government-supported organizations, such as the American Plywood Assn. or the Southern Pine Inspection Bureau, will assist upon request. These organizations, as well as others, have a record of successfully responding to members' needs for both trainers and in-process troubleshooters.

Staged Start-Up

The start-up then begins. It is usually a staged start-up beginning with the log yard and progressing through the individual departments. The more successful start-ups will feature abbreviated shift or machine schedules.

For example, if three lathes are installed, only one may be operated on day shift until each of the three is successfully debugged. Hourly employees may be doubled up on a job, with trainers covering a relatively small area. These procedures are particularly important as the industry shifts to costly high-technology systems.

Communication and Progress Tracking

The manager and the supervisors follow a clearly delineated line and scope of authority, with plans, procedures and other communication carefully documented. Communication should be concise and specific, specific enough to identify who does what by when with what results at a specific assigned area.

Daily staff meetings will supplement and enhance the ongoing verbal communication on the floor and the written confirmation that follows. The task is to have all participants "sing off the same sheet of music." Immediate feedback on results should occur as the start-up proceeds. This may include interim reports available from computerized process control centers; it will include the daily results, as with a normal production report. A formal short interval scheduling system will assist the ongoing tracking of progress.

THE TURNAROUND SITUATION

The turnaround situation is not unlike the start-up; in fact, a veteran of one or more successful start-ups is well positioned for its companion, the turnaround. The following are tips on achieving a successful turnaround.

The Plan and the Numbers

Establish an operating plan that will direct the organization to its profit goals. In other words, know where you want to go and establish a plan to get there. Then operate by the numbers.

If it doesn't show up in the numbers, then it didn't happen. Measurable results should be defined in an easy-to-understand format that is structured to meet the information and feedback needs of the recipient. The manager will be interested in overall results, with access to sufficient detail. The machine operator will usually be satisfied with the production and waste figures.

The Leadership

Poor performance is usually the result of poor leadership, either at the division or plant level. The top leadership position at the plant site must be tailored for the incumbent; success must be designed into the job. Well-defined parameters, sufficient delegation of authority for the task, thoughtful say in determining team members and a well-communicated strategy are all part of that design.

The operations manager will then establish or reestablish his presence in a well-defined situation. This may entail staff changes; it also may include operational changes that are visible and long since due.

The People Resource

Find the individual employees and determine the role and capability of each. This requires legwork, extensive interviewing and thoughtful observations. Above all it requires extensive one-on-one contact with each individual hourly and salaried employee. The manager or supervisor will then understand the strengths of each and how these strengths can be utilized to achieve the goals of the operation.

The thoughtful manager will continue this regular one-on-one contact and enhance it with small group meetings conducted by the respective department supervisors. Visits by the top leadership of the operation will be a frequent occurrence at these small group meetings.

The Equipment and Facility Resource

The symptoms of poor performance are usually reflected in how the equipment and facility are maintained. Clean, orderly and well-maintained equipment and physical structures will communicate the intentions of management to provide a workplace where individuals can be successful and can feel good about what they are doing and how they are treated.

An organized work group is a key to a successful mill turnaround.

The Elephant and the Basics

"How do you eat an elephant? One bite at a time!" The task is to select the just-right bites and then proceed with the turnaround. One bite may be yield; another may be man-hour control. The bites are selected based on the importance of each to both immediate and long-term goals. Each bite will redefine the basics. The basics are outlined in the standard operating procedures combined with a lot of horse sense.

Having a plan, organizing the people resource for results and following the basics is a systematic way of ensuring ever improving results. Frequently the can-do attitude expressed by the crew will result in performance far in excess of what the equipment and facilities would normally justify. But that's the role of the leader, whether manager or supervisor: to provide the work environment where success is normal and excellence is expected.

Twenty-four
New Technology and Innovation: A Look Ahead

The years ahead will witness an accelerating sequence of changes, changes in raw materials, products, processes and the business environment. A decade ago an industry leader commented, "The way I see it, the changes that are coming in the next three decades are going to make the last three look like a long, lazy summer" (Lewis 1974, p. 3). The subsequent years have been anything but a long, lazy summer for the forest products manufacturer.

Change will continue to offer unprecedented opportunities for those operators who anticipate it and then implement innovative ideas during the process of installing new technology. A southern panel mill is an example of both anticipating change and implementing new technology.

"New ideas mean progress" is a slogan surrounding the upper portion of the corporate logo displayed prominently on the safety award jacket worn by a plywood manufacturing manager as he discussed progress in the plant. The plant was in the midst of a $5 million modernization program, a program designed to protect and increase profits as the industry and this plant emerged from another periodic down cycle. The manager described projects planned to increase yield, cope with an even smaller log, improve man-hour productivity and enhance the product mix. This mill and this company are typical business entities that will survive and prosper.

Cited a trade journal editor as he described the needs of this and other plants, "New generation technology will be absolutely essential for lumber and plywood mills to remain solvent and competitive. Those without it will be gone or gobbled up" (Knight 1981, p. 6).

"Gone or gobbled up" can be avoided. The task is to recognize the overriding issues, implement innovative ideas and new technology in response to those issues, and look ahead while anticipating other issues in a changing business environment.

THE ISSUES

The issues can be categorized into structural issues, such as price and availability of timber, and business issues, such as technical innovation, changing consumption preferences and an evolving workforce that demands a more flexible work environment and higher work satisfaction. These issues are summarized as follows.

The Timber Resource

"Going, going and gone somewhere" is a phrase that aptly describes the skyrocketing cost and reduced availability of the traditional old-growth and larger second-growth logs. Smaller logs from diverse species and geographically dispersed stands will provide an increasing share of the log furnish for the mill.

The availability of logs from public lands will continue to be locked into non-economic public policy issues. The dialog and the results of the ensuing lobbying will continue to limit the harvest and even tie up potential sales in seemingly endless litigation.

"Practically all of the high quality old-growth softwood in the East has been exhausted. While 66 percent of the softwood sawtimber in the West is in trees 19 inches DBH and larger, only 13 percent of the softwood sawtimber in the East is in trees this size. Only 36 percent of the total softwood volume is in trees 15 inches DBH and larger" (LaBau and Knight 1978, p. 5). Utilization standards are being modified to accept the changing resource.

"The day existed when the majority of sawmills and then the plywood plants in the South specified the lower limit of merchantability of logs to be 11 inches inside bark. That limit has steadily decreased to about seven inches, and even to four inches for the Chip 'N Saw mills" (Kellison and Zobel 1978, p. 41).

The producer is being forced to use a smaller DBH stem; he is also being forced to obtain more logs from each stem and then to allocate each log to the appropriate primary conversion option.

This procedure is necessary to provide a sufficient quantity of raw material to the mill. Once in the mill, automated high-technology equipment will then extract ever more primary product from the log. Fewer chips, shavings and other by-products will be the result.

Less residuals from the solid wood conversion option combined with intense competition for the smaller log will change the economics for both the solid-wood-product manufacturer and the reconstituted board producer. No less affected will be the pulp and paper manufacturer. There will be a leveling in costs between the larger and smaller logs.

Regional Manufacturing Trends

Scattered stands, diverse species, smaller stems and changing consumer preferences will continue to fragment the industry. The strong regionalization of past years will be on the decline. The successful entrepreneur will find the market, locate the nearest sustainable patch of timber and then determine what products can be made from that tree resource to satisfy the available demand. Transportation costs and customer preferences will be important considerations.

The product will be specifically designed for the end use rather than the customary adaptation of the commodity product to the need. Nowhere is this more apparent than in structural panels.

Commented an industry leader, "The structural panel marketing arena today is characterized by substantial new developments, an air of excitement, different competitive forces, and it is not without its controversy. Headlines have loudly proclaimed the demise of the plywood mill with its rapid replacement by waferboard and oriented strand board plants. The prophecy is *at best* greatly premature" (Page 1981, p. 1).

Each unit of plywood will be manufactured with fewer and smaller logs; the other structural panels will be manufactured with some veneer, plus milled leftover log segments and residuals from the lumber and plywood manufacturing process and some small logs.

The industry will be "developing performance type standards geared only to the requirements of specific end-uses and with no reference whatsoever to panel makeup or configuration—beyond its being wood-based" (Page 1980, p. 6).

What about lumber? Opportunities for further substitution for lumber appear quite limited, but it too will be tailored to specific end-use criteria.

Mechanical stress grading (MSR) will become common as the character of the log changes from the traditional sawlog. More tops, plantation wood with a high proportion of juvenile fiber and minor species will be used. High-yield manufacturing techniques will squeeze more and more lumber from the available logs. The scragg mill will give way to the high-strain bandmill. And yet the industry will be pushed to meet the available demand.

"The projected outlook for softwood lumber consumption seems particularly optimistic. . . . Consumption is expected to exceed the 40 million board foot level throughout the 1990–2030 period. . . . Projected price growth over the next two decades is about 2.1 percent annually, as supply shifts fall behind those in demand" (Adams and Haynes 1980, p. 1). Figure 24.1 identifies this forecast.

The Accelerated Pace of Obsolescence

Obsolescence of products, manufacturing processes . . . and managers will occur at an ever more rapid pace for those who fall behind. Cited an electronics engineer during a conversation on a plane: "You know . . . each new generation of programmable logic controllers is obsolete when the customer takes delivery. By then we are about ready to put a more sophisticated unit into production. It seems that each new development spurs even more progress. It's getting tougher each year to keep up!"

The forest products industry will be fertile ground for change. Waves of product innovation will occur; there will be a heavy but temporary discounting of known and established technology . . . but a gradual leveling out as each new product finds its niche in the mill and the marketplace.

Basic manufacturing skills will grow ever closer across product lines. Increasingly the shift and the expansion of basic manufacturing to the rural areas of the U.S. West and Southeast are resulting in a draw from a common resource pool.

A hubcap plant in South Carolina is just down the road from a plywood/sawmill complex; each will draw technology and labor from the same sources. A computer manufacturer in Oregon is tapping some of the same resources as the forest products plants in the same area.

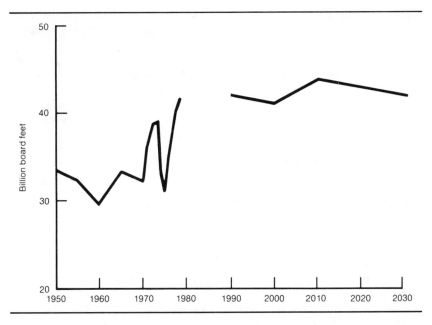

Figure 24.1. U.S. softwood lumber consumption from 1950 to 1978 and projections for 1990–2030. (Source: Adams and Haynes 1979)

Technology, labor and management skills, and a fairly uniform raw material will lend themselves to increasingly more automation. "It is getting to the point that a mill must automate or die. The market place is demanding high quality and, especially, uniformity. In return, higher production and cost savings by a mill are possible" (McCarthy 1982, p. 50).

Employees: The Business Environment and the Work Ethic

Each successful manufacturing operation is becoming more people oriented and computer/process control dominated. A contemporary Swedish professor observed: "The energy demands of work loads are comparatively low in medium-sized and large mills today; but uncomfortable postures and work movements, with a great deal of static and one-sided work loads, are frequent. Therefore, about the same relative number of workers in the mechanized medium-sized and large mills consider their work to be as physically heavy as those in the smaller, less mechanized sawmills" (Ager 1975, p. 185).

The manager's challenge is to implement methods that will yield high-quality and productive work while providing higher work satisfaction and a more positive attitude toward the company, the job and the total work environment.

Commented John Naisbitt, a noted business and social forecaster: "High tech/ high touch is a formula I use to describe the way we have responded to technology. What happens is that whenever new technology is introduced into society, there must be a counterbalancing human response . . . that is, high touch . . . or the technology is rejected. The more high tech, the more high touch" (Naisbitt 1982, p. 39).

Rotation of work and skill assignments, participation in the day-to-day organization of the work, involvement in work groups organized to solve problems, and the resulting greater frequency of contact between workers and managers will be a few of the ways increasingly used to promote and obtain higher work satisfaction. A more positive attitude to the job and the production task will be the payoffs to innovative people management methods.

The manager will increase his power to make things happen in direct proportion to his ability to respond to the employee's need for greater job satisfaction.

Issues such as the changing timber resource, manufacturing trends, the accelerated pace of obsolescence, and differing employee needs are the major concerns addressing the producer. No less important is the task of understanding and implementing innovation.

THE ROLE OF INNOVATION

Innovation occurs in response to change, and change occurs as the result of innovation. The flood of innovation will appear to engulf the manager at times. The task will be to determine the merits of each innovation. The following are five measurable attributes to look for.

Relative advantage: Is the new technology or innovation worth the effort? What tangible present or future benefits will it provide? How can these benefits be measured on the present and future operating statements?

Compatibility: How does the innovation nest into the entire system? A compatible innovation will enhance productivity and efficiency or add value to the upstream and downstream production flow.

Complexity: The complexity attribute measures the relative ease with which an innovation can be understood and implemented in the workplace. Upgrading of the work force through specialized education or recruiting will often improve upon a manufacturer's situation.

Trialability: The extent to which the innovation can be implemented a little at a time can be called its trialability. An innovation with a high degree of trialability will minimize risk and reduce actual and potential problems.

Observability: Visible results are positive results. Enthusiasm for additional yield, productivity or cost reduction efforts are usually measurably improved when the results are visible in the mill and on the statement.

In combination, the attributes will measure the total gain an innovation offers. A continuing discussion of results achieved often is perceived by some as "horn blowing," but horn blowing can be useful if it acts as a catalyst for greater enthusiasm and additional fresh, innovative ideas.

A LOOK AHEAD

The key issues define the environment in which new technology and innovation will be implemented; objectively measuring the merits of each innovation or technology application sets the stage for even more improvement. One overriding fact emerges from industry trends ... a competitive industry characterized by a cyclic auction market and sourced by a diverse and finite resource must depend on innovation and improved technology to survive.

REFERENCES

Adams, D. M., and Haynes, R. W. 1979. The Demand-Supply-Price Outlook for U.S. Timber. *Timber Demand: The Future Is Now.* Proceedings No. P-80-29. Madison, Wis.: Forest Products Research Society.

Ager, B. 1975. Improvement of Mill Working Conditions—Objectives, Benefits and Costs. *Nordic and North American Sawmill Techniques*, p. 184. San Francisco: Miller Freeman Publications.

Kellison, R. C., and Zobel, B. J. 1978. The Changing Wood Supply and Its Effect on Product. *Impact of the Changing Quality of Timber Resources.* Proceedings No. P-78-21 of the 1978 Annual Meeting, June 28, 1978, Atlanta, Ga. Madison, Wis.: Forest Products Research Society.

Knight, J. A. 1981. Editorial: Change—Challenge—Chaos? *Timber Processing Industry Issues* 6(5):6 (May).

———. 1982. Southern Timber Supplies in the 1980's. *Southern Logging Times*, June.

LaBau, V. J., and Knight, J. A. 1978. Historic Trends in the Quality of the Timber Resource Base. *Impact of the Changing Quality of Timber Resources*, p. 6. Proceedings No. P-78-21 of the 1978 Annual Meeting, June 28, 1978, Atlanta, Ga. Madison, Wis.: Forest Products Research Society.

Lewis, B. J. 1974. *Profits & Prophets: A Look into Plywood's Next Century.* Paper presented at American Plywood Assn. meeting, Portland, Oreg., June 25, 1974.

McCarthy, E. R. 1982. Mills Must Automate or Die. *Wood Based Panels International* 2(4):50 (October–December).

Muth, R. M., and Hendee, J. C. 1980. Technology Transfer and Human Behavior. *Journal of Forestry*, March:141.

Naisbitt, John. 1982. From Forced Technology to High Tech/High Touch. *Megatrends—Ten New Directions Transforming Our Lives.* New York: Warner Books.

Page, W. D. 1981. Introduction to panel discussion: Marketing Plywood and Non-Veneer Panel Products. Paper presented at the FPRS Southeastern Section annual fall meeting, "Marketing the Southern Pine Resource," 16–18 Nov., Mobile, Al.

———. 1980. New Wood-Based Structural Panel Products—An Exciting Potential for the 80s. Paper presented at *Forest Industries'* 1980 Sawmill and Panel Clinic, 1–3 Oct., Atlanta, Ga.

Glossary

Air-dried lumber: Lumber that has been dried in the open air.

Analysis of risk: Methods for determining the probability that an event will occur and the business impact of the probability.

Annual plan: A plan that identifies the operating specifics for the present or pending calendar or fiscal year; also termed *profit plan*.

Annual ring: The increment of wood generated in one year's growth; generally seen as a dense band of summer wood on a cross-section of the stem.

Attrition mill: A machine that utilizes grinding action to reduce wood particles to fibers.

Authorization for expenditure (AFE): A formal capital spending request that identifies the project, project cost and its benefits and includes other exhibits as appropriate.

Back: The lower grade side of a panel in which Appearance grade is not critical to its end use.

Band mill: A primary or secondary breakdown machine that uses band saws to reduce the log into smaller sizes; a machine that breaks down cants or lumber developing from the primary unit into their final width and thickness subsequent to sorting at the green chain or lumber sorters.

Blender: A process machine used to mix incoming materials (binder and adhesives with particles or fibers) properly before they are formed into a mat.

Block scheduling: A scheduling system that combines like products from a number of orders and runs these blocks through the production process. Mixed orders are assembled from the finished goods developed.

Board: A piece of lumber less than 2 in. in nominal thickness and 1 in. or more in width.

Board foot (bd ft): Unit of lumber measurement: 1 ft long, 1 ft wide and 1 in. thick.

Board measure (BM): Term indicating that a board foot is the unit of measurement used; often cited in thousands (MBM).

Bucking: Cross-cutting the fallen tree into predetermined lengths.

Business unit: A stand-alone business entity within the firm.

By-products: Joint products that have minor sales value compared with that of the major product or products; the residual product produced as an outgrowth of a primary manufacturing process.

Cant: A log or log segment that has been cut (slabbed) on one or more sides to square it prior to manufacture into lumber or other products.

Capital budgeting: A planning activity that identifies projects and capital requirements for replacement of or addition to the capital goods of the firm.

Carriage: A frame (on which are mounted the head blocks, setworks and other mechanisms for holding a log) that travels on tracks as it is fed through or to the sawline. The carriage is usually activated by a steam cylinder cable feed or rack-and-pinion device that propels it back and forth past the saw.

Cathode ray tube (CRT): A readout display frequently used in computer applications.

Caul plate: A flat metal sheet on which a plywood panel or a reconstituted-board mat is conveyed and pressed.

Certainty equivalent: A sophisticated method used to determine the sensitivity of the overall return on a project to assumed changes in revenues, benefits and costs. This method is well adapted to computer programming.

Chipping canter: A primary breakdown machine that makes cants from whole logs using chipping heads rather than saws.

Clipper, veneer: A plywood machine used to cut veneer ribbons or sheets into specified widths.

Cold-decking: The systematic stacking for storage, sometimes for extended periods of time, of incoming logs at the log yard at a mill or other log inventory site.

Commercial standard: *See* Product standard.

Commodity product: Certain lumber, plywood and other forest products that are manufactured under common trade specifications and sold in bulk quantities.

Composite panel: A panel constructed with softwood-veneer outer plies permanently bonded to a reconstituted wood core.

Computer: An automatic electronic machine that performs systematic calculations.

Construction lumber: Lumber that is suitable for light construction.

Conversion process: The process of manufacturing the tree into products.

Conversion return: A value measurement system that assigns the value of the end product less operating costs; quite similar to the term *return to log.*

Conversion unit: A single manufacturing facility that produces a specific product.

Cord: A measurement unit for small logs that is an equivalent in volume to 8 ft long, 4 ft wide and 4 ft high containing 128 ft^3.

Core, peeler: The inner portion of the log that remains on the lathe when the block-peeling process has been completed.

Core, plywood: The cross-banding grain direction, which runs perpendicular to that of the outer plies.

Core, reconstituted panel: The inner portion of the panel.

Core business: The main-stay commercial venture of a business, such as lumber manufacturing, timber growing or paper making.

Cost center: A segment of activity or area of responsibility for which costs are accumulated.

Critical path: The longest path of sequential events through a network model for which there is no slack time.

Critical spare parts: Parts or components that would cause substantial curtailment of a primary production line if the existing part became defective.

Critical spares program (CSP): A program that seeks to minimize the stores inventory while ensuring the timely availability of critical spare parts.

Cubic measures: Methods used to determine the volume of a log, usually expressed in cubic feet or cubic meters.

Cubic ratio: Square feet of veneer or lumber per net cubic log scale.

Cull: A tree or log that is less than one-third usable for lumber or plywood due to decay or other defects.

Cut-to-length (CTL) log processing system: A simple process system that transports cut-to-length logs into the mill conversion process.

Debarking: Manual or mechanical removal of bark from the logs prior to further processing into products.

Decision tree: A line drawing used to represent the flow or alternatives and states of nature and usually indicating the probabilities of each occurring.

Defensive capital projects: *See* Sustaining capital projects.

Diameter inside bark (dib): Diameter of a standing tree or log minus the estimated or actual thickness of the bark.

Dimension: Lumber that is from 2 in. in nominal thickness up to but not including 5 in. thick and that is 2 in. or more in nominal width.

Direct-coupled system: A veneer transport system that feeds veneer directly from the lathe to the green clipper, usually over a single tray or transport deck; also called *close-coupled system.*

Diversification: Expanding the variety of products offered from a single manufacturing process; expanding a business into different business segments or unrelated businesses.

Diversified entrant: A business entrant that uses forest products as a commercial venture (1) to smooth out the flow of earnings when combined with existing businesses or (2) as a stand-alone business to enhance company earnings.

Dryer, veneer: A mechanical dryer that conveys veneer through a heated medium on rollers, belts, cable or wire mesh. Veneer dryer controls are used to regulate conveyor speed and conditions in the drying medium.

Drying: The process of removing excess moisture from lumber, veneer, wood fibers or wood particles during the manufacturing process.

Dry kiln: An enclosed lumber-drying chamber in which temperature, humidity and ventilation are controlled for the purpose of drying to a specific moisture content.

Econometric modeling: The process of constructing business models by applying statistical methods to the study of economic data and problems.

Economic lot order (ELO): The just-right quantity by which a production process can achieve optimum cost, highest quality and timely shipments.

Edge-gluing: A process of joining two or more pieces of lumber or veneer together to form a wider width.

Edger: A downstream secondary breakdown unit used to reduce cants by squaring the edges and ripping them into lumber.

Electronic scaling: Measuring volume electronically through the use of scanners and other electronic devices.

End-dogging: A self-descriptive means of securing the log as it proceeds through the primary breakdown unit in a lumber operation.

Equilibrium moisture content (EMC): The point at which the moisture content of wood or veneer matches the moisture content of the atmosphere, whether indoors or outdoors.

Exception reporting: A written or verbal method of reporting significant variances, either positive or negative.

Face: The surface of the panel on which the grade or appearance quality is chiefly evaluated. When both surfaces are Appearance-grade quality, both are described as faces.

Falldown: A product not up to a particular grade.

Fee timber: Timber owned by the firm; term derived from the legal phrase "fee simple."

Felling: The process of cutting down a tree.

Fiberboard, medium-density: A dry-formed panel product made from wood fibers combined with a synthetic resin or other suitable binder and then compressed under heat and pressure to a density of 31 lb and more per ft³ in a hot press.

Fixed cost: A cost that, given a period of time and range of activities, does not change in total but becomes progressively smaller on a per-unit basis as volume increases.

Flaker: A mechanical device used to mill infeeding log segments into thin wood flakes suitable for use as furnish for reconstituted board products such as waferboard.

Flow process chart: A chart that visually identifies the flow of materials or people activities through a production or clerical process.

Furnish: Wood flakes, fiber or particles used as raw material in the manufacture of reconstituted board.

Gantt chart: A visual management control device in the form of a linear calendar on which future time is spread horizontally and work to be done is indicated vertically, with suitable divisions and subdivisions of time (such as months, weeks, days or hours).

Generative capital project: A capital project that is intended to improve or expand an existing business and/or businesses; also termed *offensive capital project.*

Grade: Designation of wood quality; applied to logs, lumber, boards and in-process product components. When applied to logs, grade describes the physical and quality characteristics in the various species based on official scaling and grading rules for a specific timber region; includes permissible and nonpermissible defects and minimum diameters and lengths within each grade.

Grade sawing: A sawing technique used at the headrig to cut the log for the greatest possible value recovery.

Grade stamp: An official impression on wood products indicating grade and certain other characteristics.

Grading rules: Established criteria used to evaluate varying pieces of lumber or plywood in terms of strength, appearance and suitability for the designated end use.

Green end: A manufacturing facility or portion of a larger facility that produces green veneer.

Gross feasibility analysis: A technique that broadly defines resources and benefits using discounted cash flow to determine the worth of the latter.

Group-timing technique (GTT): A work measurement technique that enables one observer using a stopwatch to make a detailed elemental time study of the multiple activities of 2 to 15 employees or machines at the same time.

Hardboard: A panel product made primarily from refined or partly refined wood fibers that are interfelted and then consolidated using a combination of pressure, heat and moisture.

Hardware: The physical components or apparatus of a computer.

High-ball: A descriptive term for a smooth and efficient operation that works at a high rate of speed.

High-yield forestry: A system of land and timber management that seeks to return the greatest net value from trees as a renewable resource.

High-yield mill management: A concept, complete with management principles, that seeks "more primary product with fewer logs and more primary product with less labor."

Horizontal integration: The process of expanding the business along related lines to maximize alternate uses of the tree. *See also* Vertical integration.

Hot press: A major piece of panel plant equipment that through heat and pressure, bonds veneer and/or reconstituted material into a panel.

Hurdle rate: A minimum accepted return on investment, usually used as a yardstick.

Industrial engineering: The process of investigating and determining design improvements within integrated systems of men, materials and equipment; draws upon specialized knowledge and skills to predict and evaluate the results of such systems.

Internal rate of return: A discounted cash flow method, defined as an interest rate, that discounts the present value of the expected cash inflows of a project compared to the cost of the investment outlay.

Iteration: A step-by-step, formalized and repetitive solution-finding process typical of linear programs.

Jib boom crane: A crane type having a counterbalanced boom that requires no outside support when functioning in the log yard.

Just in time: An inventory management system—recently popularized by the Japanese—that seeks to minimize in-process inventory.

Kerf: The actual width of the saw cut as the log is being sawn.

Keyboard (computer): A typewriter-like input device used to enter information into a computer-based system.

Labor profile: A ratio of product output per unit of labor input; usually expressed as production per man-hour or as man-hours per unit of production.

Lathe, rotary: A machine that rotates a peeler block against a full-length knife at a predetermined rate, developing a continuous ribbon of veneer of uniform thickness.

Leverage point: A point in the process or business at which a modest effort at improvement will reap disproportionately large benefits.

Limbing: Removal of limbs from a fallen tree before further processing.

Linear positioner: Hydraulically activated device used to precisely position a machine or machine component during the manufacturing process; typically used on carriage knees in a sawmill or a lathe carriage in a plywood plant.

Linear program: A computer technique that solves simultaneous linear equations using repetitive or iterative routines.

Line of balance analysis: A method used to determine the relative completion of all activities at a given point in time.

Loader: A wheeled machine used to load or unload logs.

Log boom crane: A crane used to unload and transport logs into storage or onto a mill deck. This crane type is usually counterbalanced by support legs or support structures.

Log-length processing system: A log processing system that bucks precut logs into predetermined lengths such as plywood blocks or more desirable sawlog lengths.

Log rule ratio: Square feet of veneer or lumber per board foot of net log scale.

Log rules: Methods and the results of these methods used to determine the net yield of a log, usually expressed in board feet. Scribner, Doyle and International ¼-in. are a few of the more common rules.

Log value analysis: The process of determining the value of a log within the constraints of available conversion options and sales potential.

Lumber recovery factor (LRF): The ratio of rough green lumber in board feet divided by a given cubic input of green logs, usually expressed as board feet per cubic feet (bd ft/ft³).

Machine stress rated (MSR) lumber: Lumber that has been nondestructively tested to indicate its structural value (such as modulus of elasticity). MSR lumber also must meet certain visual requirements.

Mainframe: A computer and its cabinets as distinguished from peripheral devices connected to it.

Man-hour efficiency: Level of productivity per man-hour.

Mat: Reconstituted board furnish loosely laid together after the addition of adhesives and other additives in preparation for being hot-pressed into a board.

Medium-density fiberboard: *See* Fiberboard, medium-density.

Methods analysis: A detailed study of the methods in use and opportunities for improvement.

Microcomputer: A small, specialized computer designed for discrete activities.

Microterminal: An input device (similar in appearance to a pocket calculator) used to enter machine commands, perform diagnostic tests and/or verify machine functions while the machine is operating; also termed *function keypad*.

Mission: The overall conceptual justification for operating a business.

Model: An abstract representation of reality.

Modulus of elasticity: A measurement that correlates the deflection of the lumber or board product and the load; one method of determining stiffness.

Moisture content: The weight of the water in wood expressed as a percentage of the weight of oven-dried wood.

Monte Carlo: The use of random occurrences from a probability distribution within a simulation model to describe an event in time.

Multicraft maintenance organization: A maintenance organization in which personnel are cross-trained between crafts, although certain individuals in the crew may have specialized skills.

Multiple activity process chart: A chart that graphically illustrates the operations performed simultaneously by two or more men, two or more machines or any combination of men and machines.

Net present value method: A method of calculating the expected return of a given project by discounting all expected future cash flow to the present using some predetermined minimum desired rate of return.

Network: A system of interconnecting computer-based process control systems to centralize and distribute data along with machine command signals; also termed *data highway*.

Network model: A scheduling model that lists planned activities, identifies the required sequence and timing of each and then constructs a planning model that optimizes available resources of time, people and material.

Nominal size: The size in which a product—either lumber or panels—is known and sold in the market; often different from the actual size.

Offensive capital project: *See* Generative capital project.

Old growth: A timber stand that has never been logged.

Operating plan: A short-term (usually for no more than 12 months), detailed business plan.

Order scheduling: A scheduling system that processes individual orders in sequence through the production flow.

Organized entrant: An entrepreneur or commercial group that is already organized as a business entity with a specific statement of purpose prior to entering as an active commercial participant into a new or enlarged forest products business that operates under a separate charter.

Oriented-strand board (OSB): A panel product made from milled wood fibers that are combined with a synthetic or other suitable binder and aligned in layers to enhance the structural strength of the panel. The resulting mat is then bonded under heat and pressure in a hot press.

Oven-dry weight: The point at which wood dried in a 150°C oven shows no further weight loss when drying is continued.

Overrun: The difference between the scale of a log and the board foot measure in that scale of the lumber obtained from the log expressed as a percentage of thousand-board-foot log input.

Panel product: Any one of a wide variety of wood-based products—such as plywood, particleboard, waferboard, oriented-strand board, hardboard or composites of two or more of the above—that are sold in sheets or panels.

Panorama: The scope and breadth of the planning effort as it relates to the current and expected business environment.

Particleboard: The generic term for a panel product made from milled wood particles or other ligno-cellulosic material that is combined with an adhesive and bonded under controlled heat and pressure.

Payback analysis: A method of calculating the time needed to recoup, in the form of cash inflow from the project, the initial dollars invested.

Peeler block: A precut log segment, usually cut in 8½ ft lengths, that is suitable for rotary peeling.

Peeling: Process of cutting veneer from a peeler block positioned on a rotary lathe.

Performance-rated panels: A specific group of structural panels (plywood, reconstituted board or composite board) that is individually designed and tested to meet specific performance requirements.

Performance standards: Standards established, usually using engineered methods, to determine the should-be performance of an individual, a machine or a machine center.

Phenolic-based glues: Synthetic petrochemical-based adhesives used in the manufacture of hot-pressed plywood or reconstituted board products.

Planing: The surfacing and machining to final dimension of rough lumber.

Plywood: A panel made of three or more layers of veneer joined together with glue or adhesive and usually laid up with the grain of adjoining plies at right angles. Usually an odd number of plies is used to secure balanced construction.

Poisson distribution: A mathematical probability function that is used with a model; represents the number of occurrences in a suitable interval of time.

Post-completion audit: A comparison of the actual results of an investment with the final benefit estimates that were used in approving the project.

Powered core drive: A lathe attachment that is designed to supplement the turning torque of the lathe spindle and also to counteract the core flexure that normally occurs at small core diameters.

Preconditioning: Various methods of preparing a peeler block (using a heating/wetting agent) for peeling.

Predetermined motion time (PDT) system: System of elemental manual motion times covering the principal human motions. Time values have been determined from scientific measurement for each significantly different variation of motion.

Preventive maintenance (PM): A maintenance management method that seeks to identify and correct actual and potential maintenance problems before they result in costly repairs and extensive unplanned downtime.

Preventive maintenance checklist: A routine checklist that is periodically used to check out a specific machine, machine component or machine center as part of a formal preventive maintenance program.

Primary breakdown: Breakdown of incoming logs into cants, lumber or veneer.

Primary converting plant: A plant designed to convert solid wood—in the form of a log, log segments or pieces—into a primary product.

Primary product: The major or chief product produced during the manufacturing process.

Process chart: A symbolic graphic representation of the specific steps in a processing activity.

Process control system: A statistical or electronic system used to identify the status of the on-going process and to establish control.

Processor: A computerized device used to manipulate vast quantities of information into a desired form.

Producer: A primary manufacturer of wood products (a single proprietor, a partnership, or a corporation) owning one or more mills or manufacturing one or more products as an overall business entity.

Productivity profile: The ratio of output to input.

Product standard: A voluntarily agreed upon standard that establishes and defines product requirements in accordance with the principal demands of the trade. Each standard encompasses the various forest products groupings and is administered by an outside third party agency under the auspices of the National Bureau of Standards of the U.S. Department of Commerce or a similar agency in other countries. Also termed *commercial standard.*

Profit center: A segment of the business that is responsible for both revenue and expenses.

Profit plan: *See* Annual plan.

Pro-forma statement: A forecasted financial statement based on current forecasts and best available information.

Program: A list of instructions prepared by a programmer or other trained individual that instructs a computer in the nature and sequence of the operations it is to perform.

Programmable logic controller (PLC): A computer-based process control device having fixed internal wiring and logic functions programmed into its memory; it scans incoming signals, refers to the program stored in memory and initiates corresponding output signals.

Quad saw: An arrangement of band saws used as a primary breakdown unit on logs of fairly uniform size.

Quality control: A system of auditing, evaluating and correcting to ensure that the product will conform to the end use requirements at the most economical cost while allowing for full customer satisfaction; a systematic way of guaranteeing that organized activities happen the way they are planned.

Random numbers: Numbers occurring without order or pattern.

Raw material: Logs or by-products such as sawdust, shavings, chips and other wood from which finished products are manufactured.

Reconstituted board products: Panel products that are produced from fiber or wood by-products as opposed to all-veneer panels.

Resawing: The remanufacture of a piece of lumber along its length by splitting it into predetermined thicknesses.

Residuals: Wood by-products of the primary manufacturing process; includes pulp chips, hog fuel and peeler cores.

Return on capital employed (ROCE): A benefit-measuring method, represented as a percentage of the total capital employed for a given project.

Return to log (RTL): A value measurement system that relates the value of the end product or products and the cost of obtaining those products to the cost of the log.

Ribbon: A continuous sheet of untrimmed veneer.

Sample scaling: Determining the volume of a number of stems, either before or during the measurement process, by sampling a representative population and using piece count and weight as primary indicators of volume.

Sanded plywood: An Appearance-grade panel that has been sized, sanded on two sides (S2S) and repaired as part of the finishing process.

Sander: Equipment (such as an abrasive drum belt or pad) used to impart smoothness to the panel surface and to reduce the panel to the prescribed thickness tolerance.

Sawmill Improvement Program (SIP): A yield-efficiency evaluation program developed by the Forest Service of the U.S. Department of Agriculture and administered by it and various state and federal entities; assists the sawmill operator in determining the should-be or could-be yield performance for his individual sawmill.

Scaling: Determining the volume of wood in a given quantity of log or logs, either individually or as a group.

Scarfing: Extending the length of veneer, lumber or a panel by joining pieces with chamfered joints or straight angle cuts. Adhesive is applied to the joining surfaces, imminent contact achieved and the bond secured, usually by heat and pressure.

Scragg mill: An arrangement of circular saws used as a primary breakdown unit for logs or peeler cores.

Sensitivity analysis: Analysis of the effect of a change in output for a given change in input.

Sensor: A device that responds to a physical stimulus and transmits a resulting impulse. Sensors may be contacting (mechanical) or noncontacting. *Mechanical sensors* utilize an arm or other mechanical device that is deflected by a passing object; the sensor arm or device is usually tied to a digital encoder. *Noncontacting optical sensors* utilize optical imaging reflection or other methods to ascertain the physical characteristics of a passing object.

Setworks: The machine mechanism that establishes the position of the saws on lumber manufacturing equipment.

Shake: A grain separation (between or through growth rings) that runs parallel to the length of the log or tree; it is usually considered to have occurred in the standing tree or during felling.

Sharp chain feed: A means of securing and feeding a log into a primary breakdown unit; uses sharp points on a driven infeed chain.

Short-interval scheduling system: A control method that identifies a planned quantity of work to be completed by a specific time using equally specific methods.

Sideboard: A board produced incidentally as an offshoot of cutting for other products such as dimension lumber.

Side-dogging: A method for securing the log prior to and during primary breakdown into cants and lumber.

Simulation: Manual or computerized technique for describing the actions of the real world in abstract mathematical terms. It involves determining through experiment the effects of changing the variables in the system by varying the flow of materials or resources through the operations or processes.

Size variation: Unintended deviation from a line of cut, either outside or inside the sawline, in a lumber manufacturing operation.

Skill audit matrix: A documentation of the maintenance skills required at each level of a specific job title such as millwright or electrician.

Smalian formula: Most commonly used formula for determining the cunit volume of a log cylinder.

Software: A detailed set of programs, procedures and related documentation associated with a computer system.

Softwoods: Coniferous trees characterized by needle-like leaves and scaling cones. Common North American species are Douglas fir, pine, hemlock, cedar, true firs, spruce and other related species.

Specialty product: An other-than-commodity product that is designed and constructed for a specific end use or a narrowly defined need of a customer.

Spreader: A machine that spreads glue on veneer immediately prior to panel assembly.

Standard operating procedures (SOP): A reference guide, specifically prepared for an individual mill, that outlines the manufacturing specifications for each point in the process.

State-of-the-art: The manufacturing technology in use at a specific time or during a specific period.

Strategic planning: An organized and detailed planning effort that covers a time span of 5 years or more.

Structural panel: A panel, usually tested and certified by an independent testing agency, designed for certain structural applications. Panels may be all veneer, composite construction or non-veneered reconstituted panels. Also termed *structural-use panel.*

Stumpage: Standing timber.

Sustaining capital projects: Capital projects that are intended to maintain the company's position in the business; also termed *defensive capital projects.*

Systems engineering: A structured and logically arranged method of shortening the time lag between the inception of an idea or discovery and its successful application.

Tactical plan: A short-term (up to 12 months in duration) operating plan; usually referred to as year 1 of the strategic plan.

Thumbwheel bank: A control center at which one or more rotary devices (called thumbwheels) are used to input data or instructions to the computer-based system.

Timbers: Lumber that is at least 5 in. in dimension; also classified as beam stringers and girders.

Time study: A work measurement technique that is best suited for evaluating people-paced operations. Each job or sequence of activities is broken down into elements; individual elements are timed separately as part of the whole. The resulting information is then examined to determine a more efficient method or methods.

Tomography: A concept currently used in medical diagnosis and now employed in sophisticated sensors to locate defects within the log and determine the optimum cutting practices for lumber manufacturing utilizing various cutting practices.

Topping: Removal of the unmerchantable top from the merchantable lower portion of the tree prior to transporting it to the mill.

Value engineering: The systematic use of techniques that identify a required function, establish the value for that function, and provide methods for obtaining that function at the lowest overall cost.

Value management: A management process that identifies opportunities and manages solutions that will yield the highest net return from the manufacturing process.

Variable cost: A cost that is uniform per unit of volume but that changes in direct proportion to changes in total volume; also termed a *direct cost.*

Variance: Deviation of actual results from expected or budgeted results.

Veneer: Thin sheets of wood from which plywood is made.

Veneer preparation equipment: Equipment that is utilized to process dry veneer subsequent to downstream assembly into products or before sale.

Veneer recovery factor (VRF): Square feet of veneer (⅜-in. basis) per cubic foot of gross log volume.

Vertical integration: Expanding a business on the vertical axis from timberlands to finished product: for example, a timberlands owner expanding into wood products manufacturing or a wood products manufacturer expanding into sales and distribution.

Waferboard: A panel product made from milled and dried wood flakes of either softwood or hardwood species (singly or in combination) that are combined with a synthetic adhesive and then formed into a flat panel under controlled heat and pressure in a hot press.

Waferizer: A machine that reduces log segments into wafers suitable for use in manufacturing waferboard.

Waiting line (queuing) model: A mathematical model that seeks to minimize the total cost within a line (or queue) being processed through one or more serving systems.

Wane: A scant area on lumber or veneer caused by the contour of the log from which it was cut.

Weight scaling: Determining the scale of incoming logs based on their weight and a predetermined formula or table.

Wholesaler: A middleman who buys from the manufacturer and sells to retailers, other middlemen or others.

Work order: A form that identifies and describes a maintenance need, the work to be done and the priority of the work.

Work sampling: A time study technique consisting of regular or randomly spaced instantaneous observations of work activity or delays over a total observation period. The objective of a study is usually to determine the actual percentage of time spent on the observed activities or delays.

Work simplification: A method for improving the workplace through the use by supervisors and hourly employees of analytical methods to determine how a job is done, why it is done that way, and ways of improving the present method.

XY charger: An electromechanical device that determines the optimum cylinder on an incoming peeler block and then positions the block into the lathe for peeling.

Yield: Wood fiber recovered from the log in the form of primary product as related to the volume of the original log.

Yield tracking system: A management system used to determine product volume and waste at each manufacturing point in the process.

Zero-based budgeting: A planning and budgeting system that examines and justifies each expenditure as if the need for it were entirely new.

Index

About the Author

Richard F. (Dick) Baldwin, a native of Oregon, draws on nearly three decades of direct experience in the forest products industry. Son of a 25-year industry veteran, Baldwin himself went to work at the age of 18 as a tape machine offbearer for Cascade Plywood Corp., Lebanon, Oregon. His experience has included assignments as foreman, plant industrial engineer, general superintendent, general manager and operations manager. At present, he is Vice President of Champion International Corp.'s southern softwood manufacturing operation, headquartered in Camden, Texas.

Baldwin is author of the book, *Plywood Manufacturing Practices*, now in its second edition. He also regularly contributes articles to industry publications, including *Forest Industries*, *Forest Products Journal*, *World Wood* and *Wood & Wood Products*. In addition to his writing, he is a sought-after speaker for industry symposia and seminars. Job assignments and journalistic research have taken him to the forest products manufacturing areas throughout North America and Scandinavia.

Baldwin is a member of the Plywood Pioneers Association, the Forest Products Research Society, Hoo Hoo International and the Texas Forestry Association. He is also an alternate member of the Industry Standards Committee. He holds a B.S. in production management from the University of Oregon at Eugene. Additional postgraduate work has been completed at the University of Oregon and Stephen F. Austin University.